Coding Theory - Recent Advances, New Perspectives and Applications

Edited by Sudhakar Radhakrishnan and Sudev Naduvath

Published in London, United Kingdom

IntechOpen

Supporting open minds since 2005

Coding Theory - Recent Advances, New Perspectives and Applications
http://dx.doi.org/10.5772/intechopen.92539
Edited by Sudhakar Radhakrishnan and Sudev Naduvath

Contributors
Sandeep Kaur Kuttal, Benjamin Riethmeier, Abim Sedhain, Mahammad Firose Shaik, M. Monica Subashini, G. Jaya Amrutha, Surdive Atamewoue Tsafack, Fitzroy Nembhard, Marco M. Carvalho, Özen Özer, Jagmohan Tanti, Md. Helal Ahmed, Sumant Pushp, Sergio Venturini

First published in London, United Kingdom, 2022 by IntechOpen
IntechOpen is the global imprint of INTECHOPEN LIMITED, registered in England and Wales, registration number: 11086078, 5 Princes Gate Court, London, SW7 2QJ, United Kingdom
Printed in Croatia

British Library Cataloguing-in-Publication Data
A catalogue record for this book is available from the British Library

Additional hard and PDF copies can be obtained from orders@intechopen.com

Coding Theory - Recent Advances, New Perspectives and Applications
Edited by Sudhakar Radhakrishnan and Sudev Naduvath
p. cm.
Print ISBN 978-1-83969-409-7
Online ISBN 978-1-83969-410-3
eBook (PDF) ISBN 978-1-83969-411-0

For EU product safety concerns:
IN TECH d.o.o., Prolaz Marije Krucifikse Kozulić 3, 51000 Rijeka, Croatia, info@intechopen.com or visit our website at intechopen.com.

Meet the editors

Dr. R. Sudhakar is a professor and head of the Department of Electronics and Communication Engineering, Dr. Mahalingam College of Engineering and Technology, Pollachi, India. He is also an associate editor for IEEE Access, from which he received the Outstanding Associate Editor Award in 2019. He is a reviewer of sixteen international journals, including *IEEE Transactions on Systems, Man, and Cybernetics: Systems, International Arab Journal of Information Technology, and International Journal of Computer and Electrical Engineering*, among others. He has published 110 papers in international, and national journals and conference proceedings. His areas of research include digital image processing, image analysis, wavelet transforms, and digital signal processing.

Dr. Naduvath is an Associate Professor of Mathematics at Christ (Deemed to be University), Bangalore, India. He has been teaching at the collegiate level since 2001 and has been a researcher in mathematics since 2011. His areas of specialization are graph theory, discrete mathematics, algebra, and number theory. Dr. Naduvath has more than 130 publications in international research journals to his credit. He has been a reviewer for many reputable publications including *Mathematical Reviews, zbMATH, MAA Reviews, and Computing Reviews.* He is a member of the editorial board of ten research journals and on the referee panel of more than fifty others.

Contents

Preface

Coding theory consists of techniques that enable the reliable delivery of digital data over unreliable communication channels. Such techniques allow coding, while decoding enables the reconstruction of the original data in many cases. These techniques have various applications in a variety of fields including computer science and telecommunication. These techniques also enrich the areas of information theory and error detection with various other real-life applications. This comprehensive book explores the latest developments, methods, approaches, and applications of coding theory in a wide variety of fields and endeavors. It is compiled with a view to provide researchers, academicians, and other interested readers with an in-depth discussion of the latest advances in the field. It consists of seven chapters that address a variety of subjects.

Chapter 1, "Abel and Euler Summation Formulas for $SBV(\mathbb{R})$ Functions" by Sergio Venturini, focuses on the natural setting for various Abel and Euler-Maclaurin summation formulas and their applications.

Chapter 2, "Algebraic Approximations to Partial Group Structures" by Özen Özer, algebraically demonstrates some structures of partial groups (Clifford Semigroup).

Chapter 3, "Dynamic HUB Selection Process in Wireless Body Area Network (WBAN)" by Mahammad Firose Shaik, M. Monica Subhashini and G. Jaya Amrutha, deals with wireless sensor nodes placed in, on, or around the human body to create wireless body area networks (WBANs), which are used for collecting physiological and vital signals from humans in real-time.

Chapter 4, "How Do Web-Active End-User Programmers Forage?" by Sandeep Kaur Kuttal, Abim Sedhain, and Benjamin Riethmeier, discusses the programming behavior of web-active end-users and informs researchers and professionals on how to create better support for the debugging process, which can be adapted by researchers to understand other aspects of programming such as implementing, reusing, and maintaining code.

Chapter 5, "A Public Key Cryptosystem Using Cyclotomic Matrices " by Md. Helal Ahmed, Jagmohan Tanti and Sumant Pushp, proposes an arithmetic approach for designing asymmetric key cryptography.

Chapter 6, "Conversational Code Analysis: The Future of Secure Coding" by Fitzroy Nembhard and Marco M. Carvalho looks at the limitations of existing approaches to secure coding and proposes a methodology that allows programmers to scan and fix vulnerabilities in program code by communicating with virtual assistants on their smart devices.

Finally, Chapter 7, "Non Classical Structures and Linear Codes" by Surdive Atamewoue Tsafack , discusses fuzzy sets and hyperstructures, which are considered non-classical structures for the construction of linear codes.

Dr. Sudhakar Radhakrishnan
Professor,
Department of Electronics and Communication Engineering,
Dr. Mahalingam College of Engineering and Technology,
Coimbatore – District, Tamilnadu, India

Dr. Sudev Naduvath
Christ University,
Bengaluru, India

Chapter 1

Abel and Euler Summation Formulas for $SBV(\mathbb{R})$ Functions

Sergio Venturini

Abstract

The purpose of this paper is to show that the natural setting for various Abel and Euler-Maclaurin summation formulas is the class of special function of bounded variation. A function of one real variable is of bounded variation if its distributional derivative is a Radom measure. Such a function decomposes uniquely as sum of three components: the first one is a convergent series of piece-wise constant function, the second one is an absolutely continuous function and the last one is the so-called singular part, that is a continuous function whose derivative vanishes almost everywhere. A function of bounded variation is special if its singular part vanishes identically. We generalize such space of special function of bounded variation to include higher order derivatives and prove that the functions of such spaces admit a Euler-Maclaurin summation formula. Such a result is obtained by deriving in this setting various integration by part formulas which generalizes various classical Abel summation formulas.

Keywords: Euler summation, Abel summation, bounded variation functions, special bounded variation functions, Radon measure

1. Introduction

Abel and the Euler-Maclaurin summation formulas are standard tool in number theory (see e.g. [1, 2]).

The space of *special functions of bounded variation* (SBV) is a particular subclass of the classical space of bounded variation functions which is the natural setting for a wide class of problems in the calculus of variations studied by Ennio De Giorgi and his school: see e.g. [3, 4].

The purpose of this paper is to show that this class of functions (and some subclasses introduced here of function of a single real variable) is the natural settings for (an extended version of) the Euler-Maclaurin formula.

Let us describe now what we prove in this paper.

In Section 2 we obtain some "integration by parts"-like formulas for functions of bounded variations which imply the various "Abel summation" techniques (Propositions (0.6), (0.7), and the relative examples) and in Section 3 we give some criterion for the absolute summability of some series obtained by sampling the values of a bounded variations function.

The last section contains the proofs of the main result of this paper (Theorem (0.1)) that we will now describe.

We denote by $C^1(\mathbb{R})$ (resp. $C^k([a,b])$), $L^1(\mathbb{R})$ and $L^\infty(\mathbb{R})$ respectively the space of continuously differentiable functions (resp. k-times differentiables functions on the closed interval $[a,b]$), the space of Lebesgue (absolutely) integrable functions and the space of essentially bounded Borel functions on $\mathbb{R}$.

Given $f:\mathbb{R}\to\mathbb{C}$ and $x\in\mathbb{R}$ we set

$$f(x^+)=\lim_{h\to 0^+} f(x+h), \tag{1}$$

$$f(x^-)=\lim_{h\to 0^-} f(x+h), \tag{2}$$

$$\delta f(x)=f(x^+)-f(x^-). \tag{3}$$

We denote by $BV(\mathbb{R})$ the space of bounded variation complex functions on $\mathbb{R}$; we refer to [5, 6] for the main properties of functions in $BV(\mathbb{R})$.

Any real function of bounded variation can be written as a difference of two non decreasing functions. It follows that if $f\in BV(\mathbb{R})$ then $f(x^+), f(x^-)$ and $\delta f(x)$ exist for each $x\in\mathbb{R}$ and the set $\{x\in\mathbb{R}|\delta f(x)\neq 0\}$ is an arbitrary at most countable subset of $\mathbb{R}$. Moreover, the derivative $f'(x)$ exists for almost all $x\in\mathbb{R}$ and $f'(x)\in L^1(\mathbb{R})$.

Let $f\in BV(\mathbb{R})$. We denote by df the unique Radon measure on $\mathbb{R}$ such that for each open interval $]a,b[\subset\mathbb{R}$

$$df(]a,b[)=f(a^-)-f(b^+). \tag{4}$$

We recall that f is *special* if for any bounded Borel function u

$$\int_{\mathbb{R}} u(x)df(x)=\int_{\mathbb{R}} u(x)f'(x)dx+\sum_{x\in\mathbb{R}} u(x)\delta f(x). \tag{5}$$

We denote by $SBV(\mathbb{R})$ the space of all special functions of bounded variation. We also say that $f\in BV_{loc}(\mathbb{R})$ (resp. $f\in SBV_{loc}(\mathbb{R})$) if for each $a,b\in\mathbb{R}$, with $a<b$ the function

$$f(x)=\begin{cases}0 & \text{if } x<a \text{ or } x>b,\\ f(x) & \text{if } a\leq x\leq b,\end{cases} \tag{6}$$

is in $BV(\mathbb{R})$ (resp. $SBV(\mathbb{R})$).

We define $SBV^n(\mathbb{R})$ inductively setting

$$SBV^1(\mathbb{R})=SBV(\mathbb{R}), \tag{7}$$

and for each integer $n>1$

$$SBV^n(\mathbb{R})=\{f\in SBV(\mathbb{R})|f'\in SBV^{n-1}(\mathbb{R})\} \tag{8}$$

We denote by B_n and $B_n(x)$, $n=1,2,\ldots$ respectively the Bernoulli numbers and the Bernoulli functions. Let us recall that

$$B_1(x)=\begin{cases}0 & \text{if } x\in\mathbb{Z},\\ x-[x]-\dfrac{1}{2} & \text{if } x\in\mathbb{R}\backslash\mathbb{Z},\end{cases} \tag{9}$$

where $[x]$ stands for the greatest integer less than or equal to x and $B_n(x)$, $n=2,3,\ldots$ are the unique continuous functions such that

$$B_n(x+1) = B_n(x), \tag{10}$$

$$B_n'(x) = nB_{n-1}(x), \tag{11}$$

$$\int_0^1 B_n(x)dx = 0. \tag{12}$$

Moreover $B_{2n+1} = 0$ for $n > 0$ and $B_n = B_n(0)$ for $n > 1$.

The main results of this paper is the following theorem.

Theorem 0.1 Let $f \in SBV^m(\mathbb{R})$, $m \geq 1$ and suppose $f, \ldots, f^{(m)} \in L^1(\mathbb{R})$. Then

$$\begin{aligned} \sum_{n \in \mathbb{Z}} \frac{f(n^+) + f(n^-)}{2} = \int_{\mathbb{R}} f(x)dx + \sum_{x \in \mathbb{R}} \sum_{k=1}^{m} \frac{(-1)^{k-1}}{k!} B_k(x)\delta f^{(k-1)}(x) \\ + \frac{(-1)^{m-1}}{m!} \int_{\mathbb{R}} B_m(x) f^{(m)}(x)dx. \end{aligned} \tag{13}$$

Remark. The sum "$\sum_{x \in \mathbb{R}}$" in the right hand side of the above "Euler-Maclaurin formula" (13) is actually a sum over the subset of the $x \in \mathbb{R}$ such that some of the terms $B_k(x)\delta f^{(k-1)}(x)$ do not vanish. We point out that such a set can be an arbitrary at most countable subset of $\mathbb{R}$.

Remark. Let p and q, $p < q$ be two integers and let f be a function of class C^m on the interval $[p, q]$. Set $f(x) = 0$ when x is outside of the interval $[p, q]$. Then the classical Euler-Maclaurin formula (see, e.g. Section 9.5 of [7])

$$\begin{aligned} \sum_{k=p}^{q-1} f(k) = \int_p^q f(x)dx + \sum_{k=1}^{m} \frac{B_k}{k!} \left(f^{(k-1)}(q) - f^{(k-1)}(p) \right) \\ + \frac{(-1)^{m-1} B_m}{m!} \int_p^q B_m(x) f^{(m)}(x)dx, \end{aligned} \tag{14}$$

follows easily from Theorem 0.1.

Remark. Any $f \in BV(\mathbb{R})$ decomposes uniquely as $f = f_1 + f_2 + f_3$, where $f_1(x)$ can be written in the form

$$f_1(x) = \sum_{n=1}^{+\infty} \varphi_n(x) \tag{15}$$

where each $\varphi_n(x)$ is a piece-wise constant function, $f_2(x)$ is an absolutely continuous function and $f_3(x)$ is a singular function, that is $f_3(x)$ is continuous and $f'_{3(x)} = 0$ for almost all $x \in \mathbb{R}$. Then $f = f_1 + f_2 + f_3$ is special if, and only if, $f_3 = 0$ and in this case, for each bounded Borel function $u(x)$,

$$\int_{\mathbb{R}} u(x)df_1(x) = \sum_{x \in \mathbb{R}} u(x)\delta f(x), \tag{16}$$

$$\int_{\mathbb{R}} u(x)df_2(x) = \int_{\mathbb{R}} u(x)f'(x)dx. \tag{17}$$

In this paper we do not need of the existence of such a decomposition.

2. Integration by parts formulas

Our starting point is the following theorem:

Theorem 0.2 Let $f, g : \mathbb{R} \to \mathbb{C}$ two complex function. Assume that $f \in BV(\mathbb{R}) \cap L^1(\mathbb{R})$ and $g \in BV_{loc}(\mathbb{R}) \cap L^\infty(\mathbb{R})$. Then

$$\int_{\mathbb{R}} f(x^+) dg(x) + \int_{\mathbb{R}} g(x^-) df(x) = 0, \tag{18}$$

$$\int_{\mathbb{R}} f(x^-) dg(x) + \int_{\mathbb{R}} g(x^+) df(x) = 0, \tag{19}$$

$$\int_{\mathbb{R}} \frac{f(x^+) + f(x^-)}{2} dg(x) + \int_{\mathbb{R}} \frac{g(x^+) + g(x^-)}{2} df(x) = 0. \tag{20}$$

Proof: Let $a, b \in \mathbb{R}$ with $a < b$. Theorem 7.5.9 of [5] yields

$$\int_{]a,b[} f(x^+) dg(x) + \int_{]a,b[} g(x^-) df(x) = f(b^-) g(b^-) - f(a^+) g(a^+), \tag{21}$$

$$\int_{]a,b[} f(x^-) dg(x) + \int_{]a,b[} g(x^+) df(x) = f(b^-) g(b^-) - f(a^+) g(a^+). \tag{22}$$

Since $f \in L^1(\mathbb{R})$ then necessarily

$$\lim_{b \to +\infty} f(b^-) = \lim_{a \to -\infty} f(a^+) = 0. \tag{23}$$

Since $g \in L^\infty(\mathbb{R})$ then $g(x^+)$ and $g(x^-)$ are bounded and we also have

$$\lim_{b \to +\infty} f(b^-) g(b^-) = \lim_{a \to -\infty} f(a^+) g(a^+) = 0. \tag{24}$$

and hence one obtains the formulas (18) and (19) taking the limits as $a \to -\infty$ and $b \to +\infty$ respectively in (21) and (22).

Formula (20) is obtained summing memberwise (18) and (19) and dividing by two. □

Next we prove:

Theorem 0.3 Let $f, g : \mathbb{R} \to \mathbb{C}$ two complex function. Assume that $f \in BV(\mathbb{R}) \cap L^1(\mathbb{R})$ and $g \in SBV_{loc}(\mathbb{R}) \cap L^\infty(\mathbb{R})$ and suppose that $g' \in L^\infty(\mathbb{R})$. Then

$$\int_{\mathbb{R}} f(x) g'(x) dx + \sum_{x \in \mathbb{R}}{}' f(x^+) \delta g(x) + \int_{\mathbb{R}} g(x^-) df(x) = 0, \tag{25}$$

$$\int_{\mathbb{R}} f(x) g'(x) dx + \sum_{x \in \mathbb{R}}{}' f(x^-) \delta g(x) + \int_{\mathbb{R}} g(x^+) df(x) = 0, \tag{26}$$

$$\int_{\mathbb{R}} f(x) g'(x) dx + \sum_{x \in \mathbb{R}}{}' \frac{f(x^+) + f(x^-)}{2} \delta g(x) + \int_{\mathbb{R}} \frac{g(x^+) + g(x^-)}{2} df(x) = 0. \tag{27}$$

where

$$\sum_{x \in \mathbb{R}}{}' := \lim_{\substack{a \to -\infty \\ b \to +\infty}} \sum_{a < x < b}. \tag{28}$$

Moreover, if the function f also is continuous then

$$\int_{\mathbb{R}} f(x)g'(x)dx + \int_{\mathbb{R}} g(x)df(x) = 0, \tag{29}$$

Proof: Given $a, b \in \mathbb{R}$, $a < b$ set

$$g(a,b,x) = \begin{cases} 0 & x \leq a, \\ g(x) & a < x < b, \\ 0 & x \geq b. \end{cases} \tag{30}$$

The function $h(x) = g(a,b,x)$ is in $SBV(\mathbb{R}) \cap L^{\infty}(\mathbb{R})$. Hence, formula (18) yields

$$\int_{\mathbb{R}} f(x^+)dh(x) + \int_{\mathbb{R}} h(x^-)df(x) = 0. \tag{31}$$

Since $h \in SBV(\mathbb{R})$ we have

$$\int_{\mathbb{R}} f(x^+)dh(x) = \int_a^b f(x^+)g'(x)dx + \sum_{x \in \mathbb{R}} f(x^+)\delta g(a,b,x). \tag{32}$$

But $f(x^+) = f(x)$ for almost all $x \in \mathbb{R}$ and hence

$$\int_{\mathbb{R}} f(x^+)dh(x) = \int_a^b f(x)g'(x)dx + \sum_{x \in \mathbb{R}} f(x^+)\delta g(a,b,x), \tag{33}$$

which combined with (31) yields

$$\int_a^b f(x)g'(x)dx + \sum_{x \in \mathbb{R}} f(x^+)\delta g(a,b,x) + \int_{\mathbb{R}} g(a,b,x^-)df(x) = 0. \tag{34}$$

Using the definition of $g(a,b,x)$ we have

$$\sum_{x \in \mathbb{R}} f(x^+)\delta g(a,b,x) = f(a^+)g(a^+) + \sum_{a<x<b} f(x^+)\delta g(x), \tag{35}$$

and hence

$$\sum_{a<x<b} f(x^+)\delta g(x) = -f(a^+)g(a^+) - \int_a^b f(x)g'(x)dx - \int_{\mathbb{R}} g(a,b,x^-)df(x). \tag{36}$$

As in the proof of the previous theorem we have

$$\lim_{a \to -\infty} f(a^+)g(a^+) = 0. \tag{37}$$

Since $f \in L^1(\mathbb{R})$ and $g' \in L^{\infty}(\mathbb{R})$ then $fg' \in L^1(\mathbb{R})$ and hence

$$\lim_{\substack{a \to -\infty \\ b \to +\infty}} \int_a^b f(x)g'(x)dx = \int_{\mathbb{R}} f(x)g'(x)dx. \tag{38}$$

The Radon measure $df(x)$ is bounded and the functions $x \mapsto g(a,b,x^-)$ are equibounded with respect to a and b; by the Lebesgue dominated convergence we have

$$\lim_{\substack{a\to-\infty\\ b\to+\infty}} \int_{\mathbb{R}} g(a,b,x^-)df(x) = \int_{\mathbb{R}} g(x^-)df(x). \tag{39}$$

From (36) it follows that

$$\lim_{\substack{a\to-\infty\\ b\to+\infty}} \sum_{a<x<b} f(x^+)\delta g(x) = \sum_{x\in\mathbb{R}}{}' f(x^+)\delta g(x) = -\int_{\mathbb{R}} f(x)g'(x)dx - \int_{\mathbb{R}} g(x^-)df(x) \tag{40}$$

which is equivalent to (25).

The proof of (26) is obtained in a similar manner using (19) instead of (18), and (27) is obtained summing memberwise (25) and (26) and dividing by two.

If the function g is continuous then $g(x^+) = g(x^-) = g(x)$ for each $x \in \mathbb{R}$,

$$\sum_{x\in\mathbb{R}}{}' f(x^+)\delta g(x) = 0, \tag{41}$$

and (29) follows from, e.g., (25).

□

Example. This example shows that in the hypoteses of Theorem (0.3) the series

$$\sum_{x\in\mathbb{R}}{}' f(x^+)\delta g(x) \tag{42}$$

is not, in general, absolutely convergent. Indeed, set

$$f(x) = \begin{cases} 0 & \text{if } x \le 1/2, \\ 1/x^2 & \text{if } x > 1/2, \end{cases} \tag{43}$$

and

$$g(x) = \begin{cases} 1 & \text{if } \sqrt{2n-1} < x \le \sqrt{2n}, \ n \in \mathbb{Z}, \\ 0 & \text{if } \sqrt{2n} < x \le \sqrt{2n+1}, \ n \in \mathbb{Z}. \end{cases} \tag{44}$$

Then the integral

$$\int_{\mathbb{R}} f'(x)g(x)dx = \sum_{n=1}^{+\infty} \int_{\sqrt{2n-1}}^{\sqrt{2n}} df(x) = -\sum_{n=1}^{+\infty} \frac{1}{2n(2n-1)} \tag{45}$$

is absolutely convergent, but the series

$$\sum_{x\in\mathbb{R}}{}' f(x^+)\delta g(x) = \sum_{n=1}^{+\infty} \frac{(-1)^n}{n} \tag{46}$$

is convergent but not absolutely convergent.

We also have the following theorem.

Theorem 0.4 Let $f, g : \mathbb{R} \to \mathbb{C}$ two complex function. Assume that $f \in SBV(\mathbb{R}) \cap L^1(\mathbb{R})$ and $g \in SBV_{loc}(\mathbb{R}) \cap L^\infty(\mathbb{R})$ and suppose that $g' \in L^\infty(\mathbb{R})$. Then

$$\int_{\mathbb{R}} f'(x)g(x)dx + \int_{\mathbb{R}} f(x)g'(x)dx + \sum_{x\in\mathbb{R}} \delta f(x)g(x^+) + \sum_{x\in\mathbb{R}}{}' f(x^-)\delta g(x) = 0, \tag{47}$$

$$\int_{\mathbb{R}} f'(x)g(x)dx + \int_{\mathbb{R}} f(x)g'(x)dx + \sum_{x\in\mathbb{R}} \delta f(x)g(x^-) + {\sum_{x\in\mathbb{R}}}' f(x^+)\delta g(x) = 0, \tag{48}$$

$$\begin{aligned}\int_{\mathbb{R}} f'(x)g(x)dx + \int_{\mathbb{R}} f(x)g'(x)dx + \sum_{x\in\mathbb{R}} \frac{g(x^+)+g(x^-)}{2}\delta f(x)\\ + {\sum_{x\in\mathbb{R}}}' \frac{f(x^+)+f(x^-)}{2}\delta g(x) = 0,\end{aligned} \tag{49}$$

where

$${\sum_{x\in\mathbb{R}}}' := \lim_{\substack{a\to-\infty\\ b\to+\infty}} \sum_{a<x<b} \tag{50}$$

If the function g also is continuous then

$$\int_{\mathbb{R}} f'(x)g(x)dx + \int_{\mathbb{R}} f(x)g'(x)dx + \sum_{x\in\mathbb{R}} g(x)\delta f(x) = 0, \tag{51}$$

Proof: Let f and g be as in the theorem. By formula (26) we have

$$\int_{\mathbb{R}} f(x)g'(x)dx + {\sum_{x\in\mathbb{R}}}' f(x^-)\delta g(x) + \int_{\mathbb{R}} g(x^+)df(x) = 0. \tag{52}$$

Since $f \in SBV(\mathbb{R})$, using the fact that $g(x^+) = g(x)$ for almost all $x \in \mathbb{R}$, we obtain

$$\int_{\mathbb{R}} g(x^+)df(x) = \int_{\mathbb{R}} g(x)f'(x)dx + \sum_{x\in\mathbb{R}} g(x^+)\delta f(x). \tag{53}$$

Then (52) and (53) yield (47). Formulas (48) and (49) are obtained in a similar manner using respectively Formulas (25) and (27) instead of (26).

If the function g is continuous then $g(x^+) = g(x^-) = g(x)$ for each $x \in \mathbb{R}$,

$${\sum_{x\in\mathbb{R}}}' f(x^+)\delta g(x) = 0, \tag{54}$$

and (51) follows from, e.g., (47). □

Theorem 0.4 generalizes to high order derivatives.

Theorem 0.5 Let $f, g : \mathbb{R} \to \mathbb{C}$ two complex function. Let $m > 0$ be a positive integer. Assume that $f \in SBV^m(\mathbb{R})$ with $f, \dots, f^{(m)} \in L^1(\mathbb{R})$ and $g \in SBV^m_{loc}(\mathbb{R})$ with $g, \dots, g^{(m)} \in L^\infty(\mathbb{R})$. Then

$$\begin{aligned}(-1)^{(m-1)}\int_{\mathbb{R}} f^{(m)}(x)g(x)dx + \int_{\mathbb{R}} f(x)g^{(m)}(x)dx\\ + \sum_{x\in\mathbb{R}}\sum_{k=1}^{m}(-1)^{k-1}\delta f^{(k-1)}(x)g^{(m-k)}(x^+)\\ + {\sum_{x\in\mathbb{R}}}'\sum_{k=1}^{m}(-1)^{k-1}f^{(k-1)}(x^-)\delta g^{(m-k)}(x) = 0,\end{aligned} \tag{55}$$

$$(-1)^{(m-1)}\int_{\mathbb{R}} f^{(m)}(x)g(x)dx + \int_{\mathbb{R}} f(x)g^{(m)}(x)dx$$
$$+\sum_{x\in\mathbb{R}}\sum_{k=1}^{m}(-1)^{k-1}\delta f^{(k-1)}(x)g^{(m-k)}(x^-) \tag{56}$$
$$+\sum_{x\in\mathbb{R}}{}'\sum_{k=1}^{m}(-1)^{k-1}f^{(k-1)}(x^+)\delta g^{(m-k)}(x) = 0,$$

$$(-1)^{(m-1)}\int_{\mathbb{R}} f^{(m)}(x)g(x)dx + \int_{\mathbb{R}} f(x)g^{(m)}(x)dx$$
$$+\sum_{x\in\mathbb{R}}\sum_{k=1}^{m}(-1)^{k-1}\delta f^{(k-1)}(x)\frac{g^{(m-k)}(x^-)+g^{(m-k)}(x^+)}{2} \tag{57}$$
$$+\sum_{x\in\mathbb{R}}{}'\sum_{k=1}^{m}(-1)^{k-1}\frac{f^{(k-1)}(x^-)+f^{(k-1)}(x^+)}{2}\delta g^{(m-k)}(x) = 0,$$

Proof: We prove first the formula (55). The proof is by induction on m. When $m = 1$ (55) reduces to (47). Assume that (55) holds for $m-1$, that is

$$(-1)^{(m-2)}\int_{\mathbb{R}} f^{(m-1)}(x)g(x)dx + \int_{\mathbb{R}} f(x)g^{(m-1)}(x)dx$$
$$+\sum_{x\in\mathbb{R}}\sum_{k=1}^{m-1}(-1)^{k-1}\delta f^{(k-1)}(x)g^{(m-k-1)}(x^+) \tag{58}$$
$$+\sum_{x\in\mathbb{R}}{}'\sum_{k=1}^{m-1}(-1)^{k-1}f^{(k-1)}(x^-)\delta g^{(m-k-1)}(x) = 0.$$

Replacing f with f', k with $k+1$ and changing the sign we obtain

$$(-1)^{(m-1)}\int_{\mathbb{R}} f^{(m)}(x)g(x)dx - \int_{\mathbb{R}} f'(x)g^{(m-1)}(x)dx$$
$$+\sum_{x\in\mathbb{R}}\sum_{k=2}^{m}(-1)^{k-1}\delta f^{(k-1)}(x)g^{(m-k)}(x^+) \tag{59}$$
$$+\sum_{x\in\mathbb{R}}{}'\sum_{k=2}^{m}(-1)^{k-1}f^{(k-1)}(x^-)\delta g^{(m-k)}(x) = 0.$$

Replacing g with $g^{(m-1)}$ in (47) we obtain

$$\int_{\mathbb{R}} f'(x)g^{(m-1))}(x)dx + \int_{\mathbb{R}} f(x)g^{m}(x)dx$$
$$+\sum_{x\in\mathbb{R}}\delta f(x)g^{(m-1)}(x^+) + \sum_{x\in\mathbb{R}}{}' f(x^+)\delta g^{(m-1)}(x) = 0. \tag{60}$$

Summing (59) and (60) we obtain (55).
The proofs of (56) and (57) are similar.

□

We say that a function $f \in SBV_{loc}(\mathbb{R})$ is a *step function* if $f'(x) = 0$ for almost every $x \in \mathbb{R}$.

The following propositions are easy consequences of Theorem (0.4).

Proposition 0.6 Let $[u, v] \subset \mathbb{R}$ be a bounded closed interval and let f be an absolutely continuous function on the closed interval $[u, v]$. Let $g \in SBV_{loc}(\mathbb{R})$ be a step function. Then

$$\int_u^v f'(x)g(x)dx = f(v)g(v^-) - f(u)g(u^+) - \sum_{u<x<v} f(x)\delta g(x). \tag{61}$$

Proof: First we extend the functions f as zero outside of the interval $[u, v]$. We may also assume that the function g is zero outside of a bounded open interval containing the closed interval $[u, v]$. Observe that then $f(u^+) = f(u), f(v^-) = f(v)$ and $f(u^-) = f(v^+) = 0$ and therefore $\delta f(u) = f(u)$, $\delta f(v) = -f(v)$ and $\delta f(x) = 0$ for $x \neq u, v$. By (47), we have

$$\int_{\mathbb{R}} f'(x)g(x)dx + \int_{\mathbb{R}} f(x)g'(x)dx + \sum_{x\in\mathbb{R}}{}' f(x^+)\delta g(x) + \sum_{x\in\mathbb{R}} g(x^-)\delta f(x) = 0. \tag{62}$$

Since g is a step function then $g'(x) = 0$ for almost all $x \in \mathbb{R}$ and hence it follows that

$$\int_{\mathbb{R}} f'(x)g(x)dx = -\sum_{x\in\mathbb{R}}{}' f(x^+)\delta g(x) - \sum_{x\in\mathbb{R}} g(x^-)\delta f(x). \tag{63}$$

The function f by construction has compact support, and hence, as $f(v^+) = 0$, we have

$$\begin{aligned}\sum_{x\in\mathbb{R}}{}' f(x^+)\delta g(x) &= f(u^+)(g(u^+) - g(u^-)) + \sum_{u<x<v} f(x^+)\delta g(x) \\ &= f(u)g(u^+) - f(u)g(u^-) + \sum_{u<x<v} f(x^+)\delta g(x),\end{aligned} \tag{64}$$

and

$$\sum_{x\in\mathbb{R}} g(x^-)\delta f(x) = g(u^-)\delta f(u) + g(v^-)\delta f(v) = f(u)g(u^-) - f(v)g(v^-). \tag{65}$$

Summing memberwise the last two formulas we obtain

$$\begin{aligned}\sum_{x\in\mathbb{R}}{}' f(x^+)\delta g(x) + \sum_{x\in\mathbb{R}} g(x^-)\delta f(x) &= -f(v)g(v^-) + f(u)g(u^+) \\ &+ \sum_{u<x<v} f(x^+)\delta g(x),\end{aligned} \tag{66}$$

as desired. □

Proposition 0.7 Let $f, g \in SBV_{loc}(\mathbb{R})$ be two step function. Let $[u, v] \subset \mathbb{R}$ be a bounded closed interval. Then

$$\sum_{u<x<v} g(x^+)\delta f(x) = f(v^-)g(v^-) - f(u^+)g(u^+) - \sum_{u<x<v} f(x^-)\delta g(x). \tag{67}$$

Proof: Set both the functions f and g to zero outside the closed interval $[u, v]$. Then formula (47) yields

$$\sum_{x\in\mathbb{R}} f(x^+)\delta g(x) + \sum_{x\in\mathbb{R}} g(x^-)\delta f(x) = 0. \tag{68}$$

But then

$$\sum_{x\in\mathbb{R}} f(x^+)\delta g(x) = f(u^+)g(u^+) + \sum_{u<x<v} f(x^+)\delta g(x), \tag{69}$$

and

$$\sum_{x\in\mathbb{R}} g(x^-)\delta f(x) = -f(v^-)g(v^-) + \sum_{u<x<v} g(x^-)\delta f(x); \tag{70}$$

hence

$$f(u^+)g(u^+) + \sum_{u<x<v} f(x^+)\delta g(x) - f(v^-)g(v^-) + \sum_{u<x<v} g(x^-)\delta f(x) = 0, \tag{71}$$

which is equivalent to (67). □

Example 1. (Abel summation I) Let (a_n), $n \in \mathbb{Z}$ be a sequence of complex numbers such that $a_n = 0$ for $n << 0$. Then the function

$$A(x) = \sum_{n<x} a_n \tag{72}$$

is a step function in $SBV_{loc}(\mathbb{R})$. If $f \in C^1[u,v]$ then Proposition (0.6) yields

$$\int_u^v f'(x)A(x)dx = f(v)A(v^-) - f(u)A(u^+) - \sum_{u<n<v} f(n)a_n. \tag{73}$$

Example 2. (Abel summation II) Let (a_n), (b_n), $n \in \mathbb{Z}$ be two sequence of complex numbers. Let $f, g \to \mathbb{C}$ be defined respectively setting $f(x) = a_n$ and $g(x) = b_n$ when $n \leq x < n+1$, $n \in \mathbb{Z}$. Clearly $f, g \in SBV_{loc}(\mathbb{R})$ and they are two step functions. Let be given two integers p and q, $p < q$. Set $u = p$ and $v = q+1$. Then it is easy to show that

$$\sum_{u<x<v} g(x^+)\delta f(x) = \sum_{n=p+1}^{q} b_n(a_n - a_{n-1}) \tag{74}$$

and

$$\sum_{u<x<v} f(x^-)\delta g(x) = \sum_{n=p+1}^{q} a_{n-1}(b_n - b_{n-1}); \tag{75}$$

hence, Proposition (0.7) yields

$$\sum_{n=p+1}^{q} b_n(a_n - a_{n-1}) = a_q b_q - a_p b_p - \sum_{n=p+1}^{q} a_{n-1}(b_n - b_{n-1}). \tag{76}$$

3. Sampling estimates

In this section we give some conditions which ensures the absolute convergence of series of the form $\sum_{x\in E}(f(x^-) + f(x^+))/2$ where f is a function absolutely integrable of bounded variation and E is a countable subset of $\mathbb{R}$.

The basic estimate is given in the following lemma.

Lemma 0.8 Let $A \subset \mathbb{R}$ be an open subset and let $F \subset A$ be a finite subset of A. Assume that there exist $a > 0$ such that

$$\begin{aligned} &x_1, x_2 \in F, \ \ x_1 \neq x_2 \Rightarrow |x_1 - x_2| \geq a, \\ &x \in F, \ \ y \in \mathbb{R} \backslash A \Rightarrow |x - y| \geq a/2. \end{aligned} \tag{77}$$

Then, for any complex function$f \in BV(\mathbb{R}) \cap L^1(\mathbb{R})$ we have

$$\left| \sum_{x \in F} \frac{f(x^-) + f(x^+)}{2} \right| \leq \frac{1}{a} \int_A |f(x)| dx + \frac{1}{2} \int_A |df|(x) \tag{78}$$

Proof: Let define

$$g(x) = \begin{cases} 0, & \text{if } \ x < -1/2 \ \text{ or } \ x = 0 \ \text{ or } \ x \geq 1/2, \\ x + 1/2, & \text{if } \ -1/2 \leq x < 0, \\ x - 1/2, & \text{if } \ 0 \leq x < 1/2, \end{cases} \tag{79}$$

and set

$$G(x) = \sum_{y \in F} g\left(\frac{x - y}{a}\right). \tag{80}$$

For each $x \in \mathbb{R}$ we have

$$\frac{G(x^-) + G(x^+)}{2} = G(x) \tag{81}$$

By Eq. (27)

$$-\sum_{x \in \mathbb{R}}{}' \frac{f(x^+) + f(x^-)}{2} \delta G(x) = \int_{\mathbb{R}} f(x) G'(x) dx + \int_{\mathbb{R}} G(x) df(x). \tag{82}$$

We also have

$$\delta G(x) = \begin{cases} -1 & \text{if } \ x \in F, \\ 0 & \text{if } \ x \in \mathbb{R} \backslash F, \end{cases} \tag{83}$$

which implies

$$-\sum_{x \in \mathbb{R}}{}' \frac{f(x^+) + f(x^-)}{2} \delta G(x) = \sum_{x \in F} \frac{f(x^-) + f(x^+)}{2}. \tag{84}$$

Set

$$E = \bigcup_{x \in F}]x - a, x + a[. \tag{85}$$

Then $F \subset E \subset A$ and

$$G'(x) = \begin{cases} 1/a & \text{if } \ x \in E, \\ 0 & \text{if } \ x \in \mathbb{R} \backslash E, \end{cases} \tag{86}$$

and hence

$$\int_{\mathbb{R}} f(x)G'(x)dx = \frac{1}{a}\int_E f(x)dx. \tag{87}$$

Moreover we have $G(x) = 0$ if $x \in \mathbb{R}\backslash E$ and hence

$$\begin{aligned} \sum_{x \in F} \frac{f(x^-) + f(x^+)}{2} &= \int_{\mathbb{R}} f(x)G'(x)dx + \int_{\mathbb{R}} G(x)df(x) \\ &= \frac{1}{a}\int_E f(x)dx + \int_E G(x)df(x). \end{aligned} \tag{88}$$

Taking modules, and observing that $|G(x)| \le 1/2$ for each $x \in E$, we obtain

$$\begin{aligned} \left|\sum_{x \in F} \frac{f(x^-) + f(x^+)}{2}\right| &\le \frac{1}{a}\int_E |f(x)|dx + \int_E |G(x)||df|(x). \\ &\le \frac{1}{a}\int_A |f(x)|dx + \frac{1}{2}\int_A |df|(x), \end{aligned} \tag{89}$$

as required.

Corollary 0.9 Let $f \in BV(\mathbb{R}) \cap L^1(\mathbb{R})$ and let $E \subset R$ be a countable subset. If there exists a real constant $a > 0$ such that for each pair of distinct $x_1, x_2 \in E$ we have $|x_1 - x_2| \ge a$ then

$$\sum_{x \in E} \left|\frac{f(x^-) + f(x^+)}{2}\right| < +\infty. \tag{90}$$

Proof: It suffices to choose $A = \mathbb{R}$; lemma (0.8) yields easily the assertion. □

4. Proof of Theorem 0.1

Inserting $B_m(x)$ instead of $g_m(x)$ in formula (57) of Theorem 0.5 we easily obtain

$$\begin{aligned} \sum_{n \in \mathbb{Z}}{}' \frac{f(n^+) + f(n^-)}{2} &= \int_{\mathbb{R}} f(x)dx + \sum_{x \in \mathbb{R}} \sum_{k=1}^{m} \frac{(-1)^{k-1}}{k!} B_k(x)\delta f^{(k-1)}(x) \\ &+ \frac{(-1)^{m-1}}{m!}\int_{\mathbb{R}} B_m(x) f^{(m)}(x)dx. \end{aligned} \tag{91}$$

By Corollary 0.9 it follows that

$$\sum_{n \in \mathbb{Z}}{}' \frac{f(n^+) + f(n^-)}{2} = \sum_{n \in \mathbb{Z}} \frac{f(n^+) + f(n^-)}{2} \tag{92}$$

is an absolutely convergent series, and hence Theorem 0.1 follows.

Author details

Sergio Venturini
Dipartimento di Matematica, Università di Bologna, Bologna, Italy

*Address all correspondence to: sergio.venturini@unibo.it

References

[1] Graham Everest and Thomas Ward. *An Introduction to Number Theory*. Springer-Verlag, 2005.

[2] Henri Cohen. *Number Theory Volume II: Analytic and Modern Tools*, volume 240 of *Graduate Texts in Mathematics*. Springer-Verlag, 2007.

[3] Ennio De Giorgi and Luigi Ambrosio. New functionals in the calculus of variations. *Atti Accad. Naz. Lincei Rend. Cl. Sci. Fis. Mat. Nat. (8)*, 82(2):199–210 (1989), 1988.

[4] L. Ambrosio, M. Miranda, Jr., and D. Pallara. Special functions of bounded variation in doubling metric measure spaces. In *Calculus of variations: topics from the mathematical heritage of E. De Giorgi*, volume 14 of *Quad. Mat.*, pages 1–45. Dept. Math., Seconda Univ. Napoli, Caserta, 2004.

[5] Stanislaw Lojasiewicz. *An Introduction to the Theory of Real Functions*. Wiley, 1988.

[6] Luigi Ambrosio, Nicola Fusco, and Diego Pallara. *Function of Bounded Variation and Free Discontinuity Problems*. Oxford Mathematical Monographs. Oxford University Press, 2000.

[7] Ronald L. Graham, Donald E. Knuth, and Oren Patashnik. *Concrete Mathematics*. Addison-Wesley Publishing Company, 1989.

Chapter 2

Algebraic Approximations to Partial Group Structures

Özen Özer

Abstract

In this work, we use 'Partial Group' notion and we do further investigations about partial groups. We define 'Partial Normal Subgroup' using partial conjugation criteria and we prove few results about partial normal subgroups analogous to normal groups. Also, we define congruence relation for partial groups and via this relation, we state 'The Quotient of Partial Group or Factor Group'. We give isomorphism theorems for partial groups. Explicitly, this is an analogous concept to group theory and our main is where differences partial groups from groups.

Keywords: partial group, partial normal subgroup, partial quotient group, isomorphism theorems for partial groups

1. Introduction

It is defined that a group is a set equipped with an operation described on it such that it has some properties as associated elements, an identity element, and inverse elements. Another definition can also be given as algebraic that the group is the set of all the permutations for algebraic expression's roots that displays the typical that the assembly of the permutations pertains to the set.

If questions are "how was group theory developed?, What is the importance of group theory in science or real life? investigated for the mathematical topic group theory, then we can understand easily why we work on the structures of the theory of the many types of groups.

As we know from the literature, some primary sources are determined in the development of group theory such as Algebra (Lagrange in the 17. century), Number Theory (Gauss in the 18. century) (Euler's product formula, Combinatorics, Fermat's Last Theorem, Class group, Regular primes, Burnside's lemma), Geometry (Klein, 1874), and Analysis (Lie, Poincaré, Klein in the 18. Century). It seems that three main areas have been described as Number Theory and Algebra (Galois theory, equation with degree 5, Class field theory), Geometry (Torus, Elliptic curves, Toric varieties, Resolution of singularities), and analysis in mathematics. Topology (((co)homology groups, homotopy groups) and algebraic part of it (Eilenberg–MacLane spaces, Torsion subgroups, Topological spaces), the Theory of Manifolds (manifolds with a metric), Algebra, Dynamical systems, Engineering (to create digital holograms), Combinatorial Number Theory, Mathematical Logic, Geometry in Riemannian Space, and Lie Algebra also belongs these three subjects.

Group theory is used not just in mathematics but also in computer science, physics, chemistry, engineering, and other sciences. Especially symmetry has a big potential property in the group theory. That is why it is considered as representation theory in physics. For example; mathematical works on quantum mechanics were done by von Neumann, Molecular Orbital Theory. Also, the Standard Model of particle physics, the equations of motion, or the energy eigenfunctions use group theory for their orbitals, classify crystal structures, Raman and infrared spectroscopy, circular dichroism spectroscopy, magnetic circular dichroism spectroscopy or getting periodic tables-gauge theory, the Lorentz group-the Poincaré's group in modern chemistry or physics. Also, group theory is defined as representation theory in physics. A lot of groups with prime caliber built-in cryptography for elliptic curve do a service for public-key cryptography and Diffie–Hellman key exchange takes advantage of cyclic groups (especially finite) too. Additionally, cryptographic protocols also consider infinite nonabelian groups.

We can state the applications of the group theory also in real life as follows:

1. Shopping online (we use our credit card with encryption which is obtained by group structure in RSA algorithm)

2. Music (Elementary group theory is used for the 12-periodicity in the circle of fifths in musical set theory. Transformational theory patterns musical transformations as if they are elements of a mathematical group, cyclic groups create octave and other notions, the musical actions of the dihedral groups.)

3. Medical science (to find out breast cancer) and computer science (robotics computer vision and computer graphics) and material sciences.

4. Machine learning, communication network, signal processing, etc. ...

5. Pipeline system, which is described as the Application - Business Object - Network Node Layer, is patterned and investigated by the theory of group. These systems are also related with vectors, matrices determined by group structures, and so on.

Thus, tools of the group theory are useful for working on applications in many different sciences and also real life as mentioned above.

Basic and simple examples can be given for usual groups as follows:

- Vector spaces $(V, +)$ have group structures under the addition of vectors with some properties of the scalar multiplication *.

- For p primes, elements of $\mathbb{Z}_p^* \subseteq \mathbb{Z}_p$ have algebraic structures under multiplication with unit 1.

- Assume that F be a number field and m is a natural number. Then $S = SL(m, F)$ can be determined to be the set of all regular $m \times m$ matrices with logins in F. This is a usual group with $a * b$ described by the multiplication of matrices.

We can ask readers "whether or not these examples are partial group"?

S_m is demonstrated by symmetric groups such that it includes $m!$ permutations where m objects are taken from a set A. As an illustration, we can give the symmetric group S_3. Supposing that $A = \{a, b, c\}$ and S_3 contains following objects; identity element.

$$\text{Identity element} = \begin{pmatrix} a & b & c \\ a & b & c \end{pmatrix} \text{ and others are;}$$

$$\begin{pmatrix} a & b & c \\ b & a & c \end{pmatrix}, \begin{pmatrix} a & b & c \\ a & c & b \end{pmatrix}, \begin{pmatrix} a & b & c \\ c & b & a \end{pmatrix}, \begin{pmatrix} a & b & c \\ b & c & a \end{pmatrix}, \begin{pmatrix} a & b & a \\ c & a & b \end{pmatrix}$$

There are some properties in finite or infinite usual group theory. Some of them can be seen as follows:

- Each of the elements in the finite group has finite order.
- If H is a finite group, then it is satisfied for each of the subgroups of H.
- Assuming that X, Y are groups. Then, the product of them is defined by $X \times Y = \{(a,b)|a \in X, b \in Y\}$ with unit element $e := (e_X, e_Y)$, where we write e_X, e_Y are identity element and inverse elements are in the form of (a^{-1}, b^{-1}) of in X, Y, respectively.
- Suppose that X be a usual group and a in X. So, the cyclic group $\langle a \rangle$ becomes a subgroup of X. As an illustration, we can say that $(\mathbb{Z}_m, +)$ is a cyclic group that satisfies $\langle 1 \rangle = \mathbb{Z}_m$.
- If X be a usual group with order p (p is prime), X is also a cyclic group.
- X is abelian usual group iff center of X equals to X.
- Let us consider S_m and its two permutations. These are conjugate iff their cycle types are equal/same according to their ordering.
- …
- … .

If literature (briefly, references [1–47]) is investigated, then it is easily seen that partial groups are considered as topological structures more than algebraic structures. It is tried to prepare some new algebraic perspectives/approximations for the partial group. As we know, there are many algebraic infrastructures such as finite-infinite group, abelian-nonabelian group, quaternion group, symmetric group, cyclic group, simple group, free group, orbits and stabilizers of the group, Lie group, and various kinds of theorems such as Sylow theorems, Cauchy theorem, Lagrange theorem, Cayley theorem, Isomorphism theorems as well as actions of groups for usual group theory.

An effect algebra is introduced in the foundations of mechanics [1]. Furthermore, effect algebra subjects are fundamental in fuzzy probability theory [2, 3]. Also, partial group is defined by [4] and used for topological and homological investigations. A pregroup can be defined as following:

A pregroup, [5], of a set P containing an element 1, each element $p \in P$ has a unique inverse p^{-1} and to each pair of elements $p, t \in P$ there is defined at most one product $pt \in P$ so that;

a. $1 * p = p * 1 = p$ is always defined,

b. $p * p^{-1} = p^{-1} * p = 1$ is always defined,

c. If $p * t$ is defined then $t^{-1} * p^{-1}$ is defined and equal to $(p * t)^{-1}$.

d. If $r * p$ and $p * t$ are defined then either $r * (p * t)$ is defined if and only if $(r * p) * t$ is defined in which case two are equal.

e. If $q * r, r * p$ and $p * t$ are defined then either $q * (r * p)$ is defined or $r * (p * t)$ is defined.

Every pregroup is a partial group, but the converse is not true in general. In that meaning a partial group definition can be stated as follows:

A set P is a partial group in the meaning of ([6], Lemma 4.2.5) if each associated pair $(x, y) \in P\ X\ P$ there is at most one product $x.y$ so that:

1. There is an element $1 \in P$ satisfying x.1 = 1. $x = x$ for each $x \in P$.

2. For each $x \in P$ there exists an element x^{-1} so that $x.x^{-1} = x^{-1}.x = 1$

3. If $x.y = z$ is defined so is $y^{-1}.x^{-1} = z^{-1}$.

Inspiring by the groupoid and effect algebra [7] gave an alternative partial group definition as an algebraic style. They introduced partial subgroup, partial group homomorphism, etc. as an analog investigation for group theory. Moreover, readers can learn/consider a lot more structural results on the subject from others [8–47].

In this work, some (remained ones can be considered from readers using our works) fundamental results are given for partial groups. Similarities and differences have been noticed between usual groups and partial groups. Several of them also are described in this work.

Also, we do further investigations for partial groups in algebraic style. We define partial normal subgroup and give isomorphism theorems for partial groups. This work is important because it has both topological and algebraic applications. It can be expanded to rings or other algebraic structures.

2. Preliminary results

Recently, an algebraic structure named as partial group (also known as Clifford Semigroup is isomorphic to an explicit partial group of partial mappings and it is a semigroup with central idempotents) is investigated with new structures in the literature.

A partial group (Clifford semigroup) is a regular semigroup (it means if $\mathcal{M}$ is a semigroup of group G then idempotent elements of $\mathcal{M}$ exchange with H's all elements). Another definition can be given for the partial group as "A regular semigroup with its central idempotents is named by Clifford semigroup."

Additionally, several partial algebras such as partial monoid, partial ring, partial group ring, partial quasigroup etc. ... have been worked. For example, Jordan Holder Theorem of composition series is known to hold in every abelian category. The classical theory of subnormal series, refinements, and composition series in groups is extended to the class of partial groups which is known to be precisely the classes of Clifford semigroups, or equivalently semilattices of groups. Also, relations among the language theory, words, partial groups, universal group, and homology theory have been considered with an arrow diagram of the partial group.

In this chapter, we first state some basic properties of partial groups which are mentioned in several references.

Definition 2.1 [7] Suppose G^* is a nonempty set: G^* is called as a partial group if the following conditions hold for all $x, y,$ and $z \in G^*$:

(G_1) If xy, $(xy)z$, yz *and* $x(yz)$ are defined, then the equality $(xy)z = x(yz)$ is valid.

(G_2) For each $x \in G^*$, there exists an $e \in G^*$ such that xe and ex are defined and the equality $xe = ex = x$ is valid.

(G_3) For each $x \in G^*$, there exists an $x' \in G^*$ such that xx' and $x'x$ are defined and the equality $xx' = x'x = e$ is valid.

The $e \in G^*$ satisfies (G_2) is called the identity element of G^* and the $X' \in G^*$ satisfies (G_3) is called the inverse of x and denoted by x^{-1}.

Another way, we can give partial definition as follows:

Definition 2.2. Let $\mathcal{M}$ be a semigroup. It is called *a* partial group if the followings are held.

i. Every $k \in \mathcal{M}$ has a partial identity e_k

ii. Every $k \in \mathcal{M}$ has a partial inverse k^{-1}

iii. Mapping $e_{\mathcal{M}} : \mathcal{M} \to \mathcal{M}, k \mapsto e_k$ is a semigroup homomorphism

iv. Mapping $\beta : \mathcal{M} \to \mathcal{M}, k \mapsto k$ is a semigroup antihomomorphism.

Note:

i. Let $\mathcal{M}$ be a semigroup. A partial identity of k, when exists, is unique and idempotent such that denoted by e_k.

ii. Let $\mathcal{M}$ be a semigroup. A partial inverse of $k \in \mathcal{M}$, when exists, is uniquely denoted by k^{-1}.

Definition 2.3. The regular element of the $\mathcal{M}$ semigroup is defined if there exists s $\in \mathcal{M}$ such that ysy = y. Each element of $\mathcal{M}$ is regular element also $\mathcal{M}$ is named by regular semigroup.

Definition 2.4. $\mathcal{M}$ semigroup is called as completely regular semigroup for every element s $\in \mathcal{M}$ ysy = y and ys = sy are satisfied.

Note. Unions of groups give us completely regular semigroups which are named by Clifford semigroups.

Definition 2.5. Let $\mathcal{M}$ be a semigroup. Elements k and s of a semigroup $\mathcal{M}$ are said to be inverse of each other if and only if sks = s and ksk = k.

Then following theorem can be given from the literature.

Theorem 2.1. The following results are equivalent to each other for a semigroup $\mathcal{M}$:

i. $\mathcal{M}$ is a Clifford semigroup,

ii. There exists r $\in$ S such that wrw = w and wr = rw for every w $\in \mathcal{M}$,

iii. $\mathcal{M}$ is a semilattice of groups,

iv. $\mathcal{M}$ is a completely regular inverse semigroup.

Proposition 2.1. $\mathcal{M}$ is a completely regular semigroup iff there are $e_k$and k^{-1} for every $k \in \mathcal{M}$.

Proposition 2.2 [7] Every group is a partial group and every partial group which is closed under its partial group operation is a group.

Proposition 2.3. Assuming that $\mathcal{M}$ is a partial group. Then, the following are given:

- Every idempotent element in partial group $\mathcal{M}$ is its own partial identity and partial inverse,
- *Let* $k \in \mathcal{M}$.Then, $e_k^{-1} = e_k = (e_k)^{-1}$ is hold.
- Suppose that $k \in \mathcal{M}.(k^{-1})^{-1} = k$ is satisfied.

Example 2.1 [7] Following sets with the given operation, can be seen as an example to the partial groups:

1. Let $G = \{0, \pm 1, \dots, \pm n\}$ where $n \in \mathbb{Z}^+$ and + be known addition operation on Z. Then it is easily seen that G is a partial group but is not a group.

2. Let $G = \mathbb{Z}^* \cup \{\frac{1}{n} : n \in Z^+\}$where $\mathbb{Z}^* = \mathbb{Z} - \{0\}$. So it is obvious that G is a partial group but is not a group by the known multiplication on $\mathbb{R}$.

3. Let $G = [-r, r]$ where $r \in \mathbb{R}^+$ and + be known addition operations on $\mathbb{R}$. Then it is obvious G is a partial group but is not a group.

Definition 2.6 [7] Suppose G^* be a partial group, $m \in Z^{+,}$ and $a \in G^*$. If a^m is defined and m is the least integer such that $a^m = e$, the number m is called the order of a. In this case, it is called that a has a finite order element. If there does not exist an $m \in Z^+$ such that $a^m = e$, (if only $a^0 = e$); then it is called that a has infinite order. The order of a is denoted by $|a|$.

Example 2.2 [7] $G = \{1, -1, i, -i, 2i, -\frac{i}{2}\}$ with the multiplication operation on $\mathbb{C}$ is a partial group and $|i| = 4$, $|2i| = \infty$.

Definition 2.7 [7]. Suppose G^*be a partial group. $\mathbb{Z}(G^*) = \{x \in G^* | \textit{If ax and xa are defined for all } a \in A; ax = xa\}$ is called the center of G^*.

Lemma 2.1 [7] A partial group is called *centerless* if Z (G^*) is trivial i.e., consists of only the identity element. If G^* is commutative then G^* = Z (G^*).

Definition 2.8. Supposing that $\mathcal{M} = G_p$ be a partial group and $\mathbb{T} = H_p$ be a subset of $\mathcal{M}$.$\mathbb{T}$ is called by sub partial group of $\mathcal{M}$ if $\mathbb{T}$ is a sub semigroup of $\mathcal{M}$ and e_k, k^{-1} are in $\mathbb{T}$ for all $k \in \mathbb{T}$.

Especially, $\mathcal{M}$ and the set of idempotents elements of $\mathcal{M}$ are sub partial groups of $\mathcal{M}$.

Definition 2.9 [7] Let G^* be a partial group and H_* be a nonempty subset of G^*. If H^* is a partial group with the operation in G^* then H^* is called a partial subgroup of G^*.

Example 2.3 [7] In Example 2.2 the set $G^* = \{0, \pm 1, \dots, \pm n\}$ where $n \epsilon \mathbb{Z}^+$ and + be known addition operation on $\mathbb{Z}$ is a partial group and let $H_* = \{0, \pm 1, \dots, \pm k\}$ where $0 \le k \le n$ and $k \in \mathbb{Z}$. Then H^* is a partial subgroup of G^*.

Lemma 2.2 [7] Let G^* be a partial group and H^* be a nonempty subset of G^*.H^* is a partial subgroup of G^* if and only if the following conditions hold:

i. $e \in H^*$;

ii. $a^{-1} \in H^*$ for all $a \in H^*$.

Moreover, Let G^* be a partial group and let a be an element of G^* such that the elements {a^k for all $k \in \mathbb{Z}$} are defined. Denote $\{a^k; k \in \mathbb{Z}\} = \langle a \rangle$ It is clear that the

.	e	a	b	c
e	e	a	b	c
a	a	b	c	e
b	b	c	e	a
c	c	e	d	b

Table 1.
(G) is a partial group even it has not a group structure.

set $<a>$ is a partial subgroup of G^*. The partial subgroup $<a>$ of G^* is called the cyclic partial subgroup generated by a. If there exists an element a in G^* such that $<a> = G^*$, then G^* is called a cyclic partial group.

Example 2.4 [7] Let $G = \{e, a, b, c\} \subset S = \{e, a, b, c, d\}$ and "." be a partially defined operation on G as in **Table 1**.

Remark. Note that $c.b$ is undefined. Then G is not a group but it is a partial group. Additionally, in this partial group, $<a> = G$, G is the cyclic partial group. But all partial subgroups of a cyclic partial group can not be cyclic. For instance, the partial subgroup $H = \{e, a, c\}$ is not cyclic. But in group theory, if a group is cyclic, all subgroups of it are also cyclic. The partial groups are different from groups in that meaning.

Definition 2.10. Assume that $\mathcal{M} = G_p$ be a partial group and $k \in \mathcal{M}$. Then, we define $\mathcal{M}_k = \{s \in \mathcal{M} : e_k = e_s\}$.

Theorem 2.2. Suppose that $\mathcal{M}$ is a partial group and $k \in \mathcal{M}$. Then, $\mathcal{M}_k$ is a maximal subgroup $\mathcal{M}$ of which has identity e_k and $\mathcal{M} = \bigcup\{\mathcal{M}_k : k \in \mathcal{M}\}$.

Definition 1.11 [7] Let $\mathcal{M}$ and $\mathfrak{N}$ be partial groups. A function $\sigma : \mathcal{M} \longrightarrow \mathfrak{N}$ is called a partial group homomorphism if for all $a, b \in \mathcal{M}$ such that ab is defined in $\mathcal{M}$, $\sigma(a)\sigma(b)$ is defined in $\mathfrak{N}$ and

$$\sigma(ab) = \sigma(a)\sigma(b).$$

If σ is injective as a map of sets, σ is said to be a monomorphism. If σ is surjective, σ is called an epimorphism.

Definition 2.12. For a partial group homomorphism $\sigma : \mathcal{M} \longrightarrow \mathfrak{N}$ it is defined $\ker\sigma = \{k \in \mathcal{M} : \sigma(k) = e_{\sigma(k)}\}$ and $Im\sigma = \{\sigma(k) : k \in \mathcal{M}\}$. Also, $\sigma : \mathcal{M} \longrightarrow \mathfrak{N}$ is named isomorphism if it is bijective.

As a consequence of the definition the following lemmas can be given:

Definition 2.13. Suppose that $\sigma : \mathcal{M} \longrightarrow \mathfrak{N}$ be a partial group homomorphism. Then, we define $\ker\sigma = \{k \in \mathcal{M} : \sigma(k) = e_{\sigma(k)}\}$ and $Im\sigma = \{\sigma(k) : k \in \mathcal{M}\}$.

Theorem 2.3. Assuming that $\sigma : \mathcal{M} \longrightarrow \mathfrak{N}$ be a partial group homomorphism and $k \in \mathcal{M}$. Then, the following are given.

i. $\sigma(e_k) = e_{\sigma(k)}$.

ii. $\sigma(k^{-1}) = \sigma(k)^{-1}$.

iii. $\ker\sigma$ *is a subpartial group of* $\mathcal{M}$.

iv. $Im\sigma$ *is a subpartial group of* $\mathfrak{N}$.

v. $\sigma(\mathcal{M}_k)$ *is a subpartial group of* $\mathfrak{N}_{\sigma(k)}$.

vi. $\sigma^{-1}(\mathfrak{N}_{e_k})$ *is a subpartial group of* $\mathcal{M}$.

Proposition 2.4 [7] Suppose $\mathcal{M}, \mathfrak{N}$ be partial groups and $\sigma : \mathcal{M} \longrightarrow \mathfrak{N}$ be a homomorphism of partial groups. Then the following conditions are satisfied:

i. If A is a partial subgroup of $\mathcal{M}$, then $\sigma(A)$ is a partial subgroup of $\mathfrak{N}$.

ii. If B is a partial subgroup of $\mathfrak{N}$, then $\sigma^{-1}(\mathrm{B})$ is a partial subgroup of $\mathcal{M}$.

Proposition 2.5. Let $\sigma : \mathcal{M} \longrightarrow \mathfrak{N}$ be a homomorphism of partial groups. Then, it is obtained that

$$\sigma(e_k) = e_{\sigma(k)}, \sigma\left(k^{-1}\right) = (\sigma(k))^{-1} \text{ for all } k \in \mathcal{M}.$$

Definition 2.14. A sub partial group $\mathbb{T} = H_p$ of $\mathcal{M} = G_p$ is named wide if the set of idempotent elements of $\mathcal{M}$ is a subset of $\mathbb{T} = H_p$ and normal, written $\lhd\mathcal{M}$. (if it is wide and $k\mathbb{T}k^{-1} \subseteq \mathbb{T}$ for all $k \in \mathcal{M}$).It is also trivial that the set of idempotents elements of $\mathcal{M}$ is a normal subgroup of $\mathcal{M}$ and it is called the set of idempotents elements of $\mathcal{M}$ the trivial normal subpartial group of $\mathcal{M}$.

Theorem 2.4. Assuming that $\mathcal{M}$ be a partial group and $k \in \mathcal{M}$, then

i. $\mathcal{M}_k$ is a maximal subgroup of $\mathcal{M}$ with identity e_k.

ii. $\mathcal{M} = \bigcup\{\mathcal{M}_k : k \in \mathcal{M}\} = \bigcup\{\mathcal{M}_{e_k} : e_k$ *is* in the set of idempotents elements of $\mathcal{M}\}$.

Theorem 2.5. If $\mathcal{M}$ is a partial group, then the set of idempotents elements of $\mathcal{M}$ is commutative and central.

Definition 2.15. For a partial group homomorphism$\sigma : \mathcal{M} \longrightarrow \mathfrak{N}$ it is defined $\ker\sigma = \left\{k \in \mathcal{M} : \sigma(k) = e_{\sigma(k)}\right\}$ and $Im\sigma = \{\sigma(k) : k \in \mathcal{M}\}$. Also, $\sigma : \mathcal{M} \longrightarrow \mathfrak{N}$ is named isomorphism if it is bijective.

Definition 2.16. σ is named as idempotent separating if $\sigma(e_k) = \sigma(e_s)$ implies that $e_k = e_s$, where e_k, e_s are in the set of idempotent elements of $\mathcal{M}$ for a partial group homomorphism $\sigma : \mathcal{M} \longrightarrow \mathfrak{N}$.

3. Partial normal subgroups

In group theory, normal subgroup plays an important role in the classification of groups and gives lots of algebraic results. Now, we will construct an analog definition for partial groups. Throughout G_{p} will denote the partial group. In this chapter, we should notice that if G is a group, then G is a partial group with*fect*$\{e\}$. Also, [8] in a group every element has a unique inverse, but in partial groups [7] for every element $a \in G$, we have *Inv* $(a) \neq 0$ because of that reason the identity element of the group differs from the identity element of the partial group. We can continue to work under these assumptions. From here on in, we will use the notation G_{p} for partial groups.

Definition 3.1 (Partial Conjugation Criteria). Let G_{p} be a partial group, the element $\mathrm{g_p}.x_{\mathrm{p}}.\mathrm{g_p}^{-1}$ (or $\mathrm{g_p}^{-1}$.xp.$\mathrm{g_p}$) is called partial conjugate of x_{p} by $\mathrm{g_p}$ for fixed $\mathrm{g_p}, x_{\mathrm{p}} \in G_{\mathrm{p}}$.

Theorem 3.1. Let G_{p} be a partial group and Np be a partial subgroup of G_{p} then following conditions are satisfied:

i. N_{p} is normal in G_{p} if and only if for all $x_{\mathrm{p}} \in N_{\mathrm{p}}$ and $\mathrm{g_p} \in G_{\mathrm{p}}$ we have $\mathrm{g_p}^{-1}.x_{\mathrm{p}}.\mathrm{g_p} \in N_{\mathrm{p}}$

ii. N_p is normal in G_p if and only if for every element of N_p all partial conjugates of that element also lie in N_p.

Proof:

i. It comes from the definition of partial normal subgroup.

ii. (⇒:) Let N_p is partial normal subgroup in G_p. Then we need to show for every $x_p \in Np$ and fixed $g_p \in N_p$ the partial conjugates $g_p^{-1}.x_p.g_p$ lies in N_p. Since $g_p \in Np$ then $g_p \in G_p$. Also, since N_p is a partial normal subgroup of G_p $g_p^{-1}.x_p.g_p \in N_p$, this gives the proof.

(⇐:) Conversely, let for every element of N_p all partial conjugates of that element lie in Np: Then it comes directly from partial subgroup definition Conclusion(s). It is preferable to include a Conclusion(s) section which will summarize the content of the book chapter.

Remark 3.1. Any partial subgroup $H_p \subseteq G_p$ has right and left congruence (equivalence) class that cannot be the same. But if left and right congruence classes are the same (i.e., for any $x \in G_p$, $H_p.x = x.H_p$) then H_p is called as normal partial subgroup.

Theorem 3.2. Let G_p be a partial group and N_p be a partial subgroup of a partial group G_p and so the following conditions are coincided:

Proposition 3.1.

i. N_p is a partial normal subgroup of a partial group G_p.

ii. $g_p.N_p = N_p.g_p$ for all $g_p \in G_p$,

iii. $g_p.N_p.g_p^{-1} \subseteq N_p$ for all $g_p \in G_p$.

Proof:

(i) ⇒(ii) If N_p is a partial normal subgroup of G_p then it is easy that for all $x_p \in N_p$ and $g_p \in G_p$ we have $g_p.x_p.g_p^{-1} \in N_p$ and from just before the theorem $g_p.N_p.g_p^{-1} \in N_p$.

(iii)⇒(ii) Let g_p be an element of G_p: We need to see $g_p.N_p = N_p.g_p$. Assume that $x_p \in g_p.N_p$. Then $x_p = g_p.n_1$ is satisfied for $n_1 \in N_p$. Since$x_p.g_p^{-1} = g_p.n_p.g_p^{-1}$ and $g_p.N_p.g_p^{-1} \subseteq N_p$ we have $x_p g_p^{-1} \in N_p$ and so that there exists an $n_2 \in N_p$ such that $x_p.g_p^{-1} = n_2$. If we product from right with g_p then we have the $(x_p.g_p^{-1}).g_p = n_2.g_p$ equality.

Using the associativity property the equality becomes $x_p.(g_p^{-1}.g_p) = n_2.g_p$ and using identity element property and converse element property we get $x_p = n_2.g_p \in N_p.g_p$. It implies that $g_p.N_p \subseteq N_p.g_p$. In a similar way $g_p^{-1}.N_p.g_p \subseteq N_p$ and we have $N_p.g_p \subseteq g_p.N_p$. Therefore we obtain $g_p N_p = N_p.g_p$.

(ii)⇒(i) Supposing that $g_p \in G_p$ we have to prove $g_p.n_p.g_p^{-1}$ is contained in N_p for all $n_p \in N_p$. Since $g_p.N_p = N_p.g_p$, we can say that $g_p.np \in N_p.g_p$ for all $n_p \in N_p$. Then, by associativity $g_p.n_p.g_p^{-1}$ is contained in $N_p.g_p.g_p^{-1}$ and then for all $n_p \in N_p$, $g_p.n_p.g_p^{-1}$.

Lemma 3.1. The center of a partial group G_p is a partial normal subgroup of G_p.

Proof: The center of a partial group is defined as below:

$Z(G_p) = \{x_p \in G \mid \text{If for every } g_p \in G_p\ x_p.g_p \text{ and } g_p.x_p \text{ are defined, } x_p.g_p = g_p.x_p\}$.

Let $g_p \in G_p$ and $x_p \in Z(G_p)$ then we need to show $g_p.x_p.g_p^{-1}$ is contained in $Z(G_p)$. Since $x_p \in Z(G_p)$ for every $g_p \in G_p$ if $g_p.x_p$ and $x_p.g_p$ is defined then $x_p.g_p = g_p.x_p$. Using this argument $g_p.x_p.g_p^{-1} = x_p.g_p.g_p^{-1} = x_p \in Z(G_p)$ and then we have $Z(G_p)$ is a partial normal subgroup of G_p.

Proposition 3.2. Let φ: $G_p \rightarrow H_p$ be a partial group homomorphism. Then the kernel of φ is partial normal subgroup of G_p.

Proof: Let K = $Ker(\varphi)$. We know that $Ker(\varphi)$ is a partial subgroup of $G_{p.}$ Suppose that $y \in K$ and $g_p \in G_p$. Then using the fact φ is a partial group homomorphism

$$\varphi\left(g_p \cdot y_p \cdot g_p^{-1}\right) = \varphi\left(g_p\right) \cdot \varphi\left(y_p\right) \cdot \varphi\left(g_p^{-1}\right)$$
$$= \varphi\left(g_p\right) . eH_p . \varphi\left(g_p^{-1}\right)$$
$$= \varphi\left(g_p\right) . \varphi\left(g_p^{-1}\right)$$
$$= eH_p \text{we have } \varphi\left(g_p \cdot y_p \cdot g_p^{-1}\right) = eH_p$$

and we get $g_p \cdot y_p \cdot g_p^{-1} \in K$. So Ker($\varphi$) is a partial normal subgroup of Gp.

Partial normal subgroups

Theorem 3.3. Suppose G_p is a partial group and H_p is a partial subgroup of G_p.H_p is a partial normal subgroup of G_p if and only if (aH_p) (bH_p) = abH_p equality holds for all $a, b \in H_p$.

Proof:

($\Rightarrow$:) Suppose H_p is a partial normal subgroup of G_p. We need to see the equality (aH_p) (bH_p) = abH_p holds.

($\subseteq$:) Let $x \in$ (aH_p) (bH_p) then $\exists h_, h_2 \in H$ such that x = (ah_1) (bh_2). By using associativity property, we get $x = (h_1 b h_2)$. From the identity element, $x = abb^{-1}h_1bh_2$ is obtained. Considering associativity property we can write $x =$ ab $(b^{-1}h_1b)h_2$. Since H_p is a partial normal subgroup of G_p, $b^{-1}h_1b \in H_p$ Then x $\in abH_p$ and this implies that (aH_p) (bH_p) $\subseteq abH_p$.

($\supseteq$:) Conversely, let $y \in abH_p$ then there exists an $h \in H_p$ such that $y = abh$. So we can write y = abh as follows; $y = (ae)(bh) \in (aH_p)$ (bH_p). It implies that $abH_p \subseteq (aH_p)$ (bH_p). Therefore (aH_p) $(bH_p) = abH_p$

($\Leftarrow$:) Let us consider (aH_p) $(bH_p) = abH_p$ for all $a, b \in G_p$. If $h \in H_p$ and $g \in G_p$ then we must see whether or not $ghg^{-1} \in H$. Using associativity property and considering hypothesis; ghg^{-1} = $(gh)(g^{-1}e)$ $\in$ $(gH_p)(g^{-1}H_p)$ = $gg^{-1}H_p$ = H_p. This implies that $ghg^{-1} \in H_p$. So that H_p is a partial normal subgroup of G_p.

Theorem 3.4. Suppose G_p be a partial group and H, K are partial subgroups of G_p. If K is a partial normal subgroup of G_p, then the following cases are satisfied:

i. $H \cap K$ is a partial normal subgroup of H.

ii. If K and H are partial normal subgroups of G_p and $H \cap K$ = $\{e\}$ then $hk = kh$ (or, $HK = KH$) for every $h \in H$ and every $k \in K$.

Proof:

i. Since H and K are partial subgroups, $H \cap K$ is also a partial subgroup of G. By $H \cap K \subseteq K$ we can conclude that $H \cap K$ is also a partial subgroup of H. Let consider $a \in H \cap K$ and $h \in H$: Since $a, h \in H$, ha and hah^{-1} can be defined. It gives that $hah^{-1} \in$ H Also since K is a partial normal subgroup of G_p we have $hah^{-1} \in K$. It shows that $H \cap K$ is a partial normal subgroup of H.

ii. Let H be a partial normal subgroup of G_p, $H \cap K = \{e\}$ and $h \in H$ and $k \in K$. Since H is a partial normal subgroup, we know that $k.h.k^{-1} \in H$. If $h.k.h^{-1}.k^{-1} \in H$ then we get $(h.k.h^{-1}).k^{-1} \in K$ by K is a partial normal subgroup. So we have, $h.k.h^{-1}.k^{-1} \in H \cap K = \{e\}$. Then $h.k.h^{-1}.k^{-1}=e$ and so, $hk = kh$; for all $h \in H, k \in K$. Thus, we prove that $HK = KH$.

Proposition 3.3. Let G_p be a partial group and H_p be a partial subgroup of index 2 in G_p: Then H_p is a partial normal subgroup in G_p.

Proof: Let H_p be a partial subgroup of index 2 in G_p and g_p be an element of G_p. If $g_p \in H_p$, then $g_pHp = H_pg_p$ is satisfied. If g_p is not in H_p, two left cosets must be as H_p and g_pH_p. Since left cosets are disjoint we know $g_p.H_p = G_p-H_p$. Also, the right cosets are disjoint so we can write $H_p.g_p = G_p-H_p$. Thus $g_pH_p = H_pg_p$ for all $g_p \in G_p$. So H_p is normal.

Example 3.1. Let G_p be an abelian partial group. Then any subgroup of G_p is a partial normal subgroup of G_p.

Proof:

If G_p is an abelian partial group and $x_p.y_p \in G_p$ then $x_p.y_p = y_p.x_p$ for every $x_p.y_p \in G_p$. If $g_p.x_p.g_p^{-1} \in N_p$ then N_p is a partial subgroup of G_p. Using the hypothesis we get

$$\begin{aligned} g_p.x_p.g_p^{-1} &= g_p.g_p^{-1}.x_p \\ &= eG_p \\ &= x_p \in N_p \end{aligned}$$

Therefore, the partial subgroup N_p is the partial normal subgroup of G_p.

Example 3.2. Let Hp and K_p be any partial normal subgroup of G_p. Then $H_p \times K_p$ is also a partial normal subgroup of $G_p \times G_p$.

Proof:

$H_p \times K_p = \{n_p=(h_p,k_p) \mid h_p \in H_p \text{ and } k_p \in K_p\}$. We have to show that for all g_p in G_p and n_p in $H_p \times K_p$ $g_p.n_p.g_p^{-1}$ is in $H_p \times K_p$.

$$\begin{aligned} g_p.n_p &= g_p.(h_p, k_p) \\ &= \left(g_p.h_p, g_p.k_p\right) \end{aligned}$$

and

$$\begin{aligned} g_p.n_p.g_p^{-1} &= \left(g_p.h_p, g_p.k_p\right).g_p^{-1} \\ &= \left(g_p.h_p.g_p^{-1}, g_p.k_p.g_p^{-1}\right) \end{aligned}$$

and since H_p and K_p are partial normal subgroups of G_p, then $g_p.h_p.g_p^{-1} \in H_p$, and $g_p.k_p.g_p^{-1} \in K_p$ and so that $(g_p.h_p.g_p^{-1}, g_p.k_p.g_p^{-1}) \in H_p \times K_p$. Then the Cartesian product of two partial normal subgroups is also a partial normal subgroup.

Theorem 3.5. Let be a partial group and N_p be a partial normal subgroup of G_p. The congruence modulo N_p is a congruence relation for the partial group operation ".".

Proof. Let $x\Re N_p y$ denote that x and y are in the same coset, that is; $x\Re N_p y \Leftrightarrow x.N_p = y.N_p$

Let $x\Re N_p x$ and $y\Re N_p y$. To demonstrate that $\Re N_p$ is a congruence relation for ., we need to show, reflexivity, symmetry, and transitivity. These axioms are obvious from the definition of relation.

Theorem 3.6. If N_p is a partial normal subgroup of a partial group G_p and $G_p= N_p$ is the set of all cosets of N_p in G_p, then G_p/N_p is a partial group under the operation given by $(aN_p)(bN_p) = abN_p$.

Proof. Let aN_p, bN_p,$cN_p \in G_p/N_p$. We must see partial group axioms are satisfied:

(G1) If $(aN_p)(bN_p) = abN_p$. $(bN_p)(cN_p) = bcN_p$ and $a.(bcN_p)$ are defined then

$$\begin{aligned} a.(bcN_p) &= (aN_p)((bN_p)(cN_p)) \\ &= ((aN_p)((bN_p))(cN_p) \\ &= (ab)cN_p \end{aligned}$$

(G2) For any aN_p in G_p/N_p, eN_p is a candidate for identity element, i.e.,

$$\begin{aligned} (aN_p)(eN_p) &= aeN_p \\ &= aN_p \end{aligned}$$

and

$$\begin{aligned} (eN_p)(aN_p) &= eaN_p \\ &= aN_p \end{aligned}$$

(G3) Since is a partial group $a \in G_p$ has an inverse $a' \in G_p$. For every aN_p in G_p/N_p $a'N_p$ is a candidate for the inverse of aN_p.

$$\begin{aligned} (aN_p)(a'N_p) &= (a\acute{a})N_p \\ &= eN_p \\ &= N_p \end{aligned}$$

This completes the proof.

Definition 3.2. Let G_p be a partial group and Np is a partial normal subgroup of G_p, then the partial group G_p/N_p is called the quotient of the partial group or factor group of G_p by N_p.

Proposition 3.4. Let $f : G_p \longrightarrow H_p$ be a homomorphism of partial groups, then the kernel of f is a partial normal subgroup of G_p. Conversely, if N_p is a partial normal subgroup of G_p, then the map $\prod$: $G_p \longrightarrow G_p/N_p$ given by $\prod(a_p) = aN_p$ is an epimorphism with kernel N_p.

Proof: $Kerf = \{x_p \in G_p | f(x_p) = eH_p\}$, we need to show that if $x_p \in Kerf$ and $a_p \in G_p$ whether or not $a_p x_p a_p^{-1} \in Kerf$ if and only if $f(a_p x_p a_p^{-1}) = f(a_p) f(x_p) f(a_p^{-1}) = eH_p$

So, $a_p x_p a_p^{-1} \in Kerf$. Therefore, $Kerf$ is a partial normal subgroup of G_p. It is trivial that $\prod$: $G_p \longrightarrow G_p/N_p$ is surjective. $Ker(\prod) = \{x_p | \prod(x_p) = e_p N_p\} = N_p$.

Theorem 3.7. Let $f : G_p \rightarrow H_p$ is partial group homomorphism and N_p is a partial normal subgroup of G_p contained in the kernel of f, then there is exactly unique homomorphism $\bar{f}$: $G_p/N_p \longrightarrow H_p$ such that $\bar{f}(aN_p) = f(a)$ for all $a \in G_p$. Besides $Imf = Im\bar{f}$ and $ker\bar{f} = (kerf)/N_p$.$\bar{f}$ an isomorphism if and only if f is an epimorhism and $N_p = ker f$.

Proof: $f : G_p \xrightarrow{f} H_p$, $G_p \longrightarrow G_p/N_p$, $G_p/N_p \xrightarrow{\bar{f}} H_p$ diagram is commutative. If $b \in aN_p$, then $b = an_p$,$n_p \in N$ and also $f(b_p) = f(an_p) = f(a)f(n_p) = f(a)e = f(a)$, since $N_p \leq ker f$. Therefore, f has the same effect on every element of aN_p and the map $\bar{f}$: $G_p/N_p \longrightarrow H_p$ given by $\bar{f}(aN_p) = f(a)$. It is easily seen that $\bar{f}$ is a well-defined function. Now we need to prove whether or not $\bar{f}$ is a homomorphism of partial groups.

$$f(aN_p.bN_p) = \bar{f}(abN_p) = f(ab)$$

$$= f(a).f(b)$$

$$= \overline{f}(aN_p).\overline{f}(bN_p)$$

So, f is a partial group homomorphism. $Im\overline{f} = Imf$ and $aN_p \in ker\overline{f} \Leftrightarrow f(a) = e \Longleftrightarrow a \in kerf$, whence

$$ker\overline{f} = \{aN_p | a \in kerf\}$$

$$= kerf/N_p$$

$\overline{f}$ is unique since it is completely determined by f. Also, $\overline{f}$ is a partial group if and only if f is an epimorphism of partial groups $\overline{f}$ is a monomorphism if and only if for $ker\overline{f} = (ker f)/N_p ker f$ equal to N_p.

Example 3.3. In Example 2.2, it is stated that $G = \{0,\pm1,\pm2, \dots ,\pm n\}$ is a partial group with known addition operation on $\mathbb{Z}$. We can easily say that the subset $N = \{0,\pm1,\pm, \dots ,\pm n - 1\}$ of G is a partial subgroup of G. Let us show whether or not N is a partial normal subgroup. If $g_p^{-1} + n_p + g_p \in N$ for all $g_p \in G_p$ then N is a partial normal subgroup. $g_p^{-1} = g_p$ in this group so that $g_p + n_p + g_p \in N_p$ i.e. $n_p \in N$ and then N is a partial normal subgroup of G.

Theorem 3.8 (First Isomorphism Theorem for Partial Groups). Let G_p, H_p be partial groups and $f : G_p \longrightarrow H_p$ be partial group epimorphism then $G_p/Ker f$ is isomorphic to H_p.

Proof: We know that $ker f$ is a partial normal subgroup of G_p. Then $G_p/Ker f$ is defined. Let show $Ker f = K$ and g: $G_p/K \longrightarrow H_p$ mapping defined as $g((aK)(bK)) = g(abK) = f(ab)$ for every $aK, bK \in G_p/K$.

Since $g(a.bK) = f(ab) = f(a)f(b) = g(aK)g(bK)$ then g is a homomorphism. Since f is onto there exists $a \in G_p$ such that $f(a) = h$ then $aK \in G_p/K$ and $g(aK) = f(a) = h$ and g is onto. For one-to-one conditions let $aK, bK \in G_p/K$ and

$$g\,(aK) = g(bK) \text{ iff} f(a) = f(b)$$

$$f(b)^{-1}f(a) = f(b)^{-1}f(b) \text{ iff} f(b^{-1}a) = eH_p$$

$$b^{-1}a \in K \text{ or } kerf \text{ iff} aK = bK$$

Then, g is an isomorphism and $G_p/Ker f \cong H_p$.

Lemma 3.2. Let G_p be a partial group and H_p be a partial subgroup of G_p and N_p be a partial normal subgroup of G_p. If H_p is a partial normal subgroup of G_p; then H_pN_p is a partial normal subgroup of G_p.

Proof: Trivial.

Theorem 3.9 (Second Isomorphism Theorem for Partial Groups). Let G_p be a partial group. H_p be a partial subgroup of G_p and N_p be a partial normal subgroup of G_p, then the following isomorphism holds:

$$H_pN_p/N_p \cong H_p(H_p \cap N_p)$$

Proof: It can be easily seen using First Isomorphism Theorem. So, the proof is left to the reader.

Theorem 3.10 (Third Isomorphism Theorem for Partial Groups). Let G_p be a partial group and H_p,N_p partial normal subgroups of G_p with $H_p \leq N_p$ Then H_p is also a partial normal subgroup of N_p.N/H_p is a partial normal subgroup of G_p/H_p and also G_p/N_p is isomorphic to $(G_p/H_p)/(N_p/H_p)$.

Proof: Let *define* $f: G_p/H_p \longrightarrow G_p/N_p, f(aH_p) = aN_p$ First let show f is well-defined: If $aH_p = bH_p$ then we need to prove whether or not $f(aH_p) = f(bH_p)$. Since $aH_p = bH_p$ then $ab^{-1} \in H_p$ and $H_p \leq N_p, ab^{-1} \in N_p$. Since $ab^{-1} \in N_p$ then $aN_p = bN_p$ and so that $f(aH_p) = f(bH_p)$, i.e., f is well defined. f is homomorphism. And using the First Isomorphism Theorem we can conclude the result.

4. Conclusion

There are some papers such as solvable partial groups, topological structures of partial group, Transitivity Theorem-Thompson Theorem of partial groups, k-partial groups, primitive pairs of partial groups so on related with the partial group in the literature. It is known that every group is partial but the converse is not true. That is why some structures are different from each other for the usual group and partial group.

In this chapter, some structures of partial groups (Clifford Semigroup) are sought to demonstrate algebraically. At the beginning of the chapter (preliminaries section), several fundamental results of partial groups with some numerical examples are given from the literature. For example, if A and B be two usual groups such that the intersection of them is equal to {1 = e}, then the union of subgroups of A and subgroups of B is a partial group. Partial normal groups and partial quotient groups have introduced an analog of the group theory. By using them, a number of isomorphism theorems are proved for partial groups with several other ideas. All results are obtained using closely group theory as algebraic approximations. Readers also may consider/investigate other structures/properties of the partial groups different from the group as algebraically.

Conflict of interest

The author declare no conflict of interest.

Author details

Özen Özer
Department of Mathematics, Faculty of Science and Arts, Kırklareli University, Kırklareli, Turkey

*Address all correspondence to: ozenozer39@gmail.com; ozenozer@klu.edu.tr; ooozenozer@gmail.com

References

[1] Foulis D, Bennett MK. Effect algebras and unsharp quantum logics. Foundations of Physics. 1994;**24**:1331-1352

[2] Beltrametti EG, Bugajski S. A classical extensions of quantum mechanics. Journal of Physics A: Mathematical and General. (IOP Publishing Ltd). 1995;**28** (12):3329-3343

[3] Bugajski S. Fundamentals of fuzzy probability theory. International Journal of Theoretical Physics. 1996;**35**: 2229-2244

[4] Jekel S. Partial Groups, Northeastern Univeristy Representation Seminar Talk 10/08/2013. Northeastern University. DOI: 10.13140/2.1.1409.3127

[5] Stallings JR. The cohomology of pregroups. In: Gatterdam RW, Weston KW, eitors. Lecture Notes in Mathematics. Heidelberg: Springer; 1973

[6] Isaacs M. Finite Group Theory. American Mathematical Society. In: Graduate Studies in Mathematic. Vol, 92. Rhode Island, USA: Hardcover MSC; 2008. p. 350. Primary 20; Print ISBN: 978-0-8218-4344-4

[7] Ciloglu Sahin Z, Ceven Y. Generalization of groups: Partial groups. In: Emerging Applications of Differential Equations and Game Theory. IGI Global; 2020. pp. 1-12. DOI: 10.4018/978-1-7998-0134-4.ch001

[8] Bogopolski O. Introduction to Group Theory, EMS, Textbooks in Mathematics. Switzerland: European Mathematical Society; 2008. DOI: 10.4171/041

[9] Abd-Allah AM, Abdallah ME-GM. Quotient in partial groups. Delta Journal of Science. 1984;**8**(2):470-480

[10] Abd-Allah AM, Abdallah ME-GM. On Clifford semigroup. Pur. Math. Manuscr. 1988;**7**:1-17

[11] Abd-Allah AM, Aggour AI, Fathy A. Strong semilattices of topological groups. Journal of Egyptian Mathematical Society. 2016;**24**:597-602. DOI: 10.1016/j.joems.2016.03.003

[12] Abd-Allah AM, Aggour AI, Fathy A, k-partial groups, Journal of Egyptian Mathematical Society. 2017;**25**(3): 276-278. doi: 10.1016/j.joems.2017.01.008

[13] Aschbacher M. Finite Group Theory. Cambridge Studies in Advanced Mathematics. 1st ed. Vol. 10. Cambridge: Cambridge University Press; 1986

[14] Bender H. On groups with abelian Sylow 2-subgroups. Mathematische Zeitschrift. 1970;**117**:164-176

[15] Bergelson V, Blass A, Hindman N. Partition theorems for spaces of variable words. Proceedings of the London Mathematical Society. 1994;**68**:449-476

[16] Brandt W. Über eine Verallgemeinerung des Gruppengriffes. Mathematische Annalen. 1926;**96**:360-366

[17] Brown R. Topology and Groupoids. Oxon: McGraw-Hill; 2006 this updated version first published in 2006 by www. groupoids.org.uk Deganwy, United Kingdom

[18] Delizia C, Dietrich H, Moravec P, Nicotera C. Group in which every non-abelian subgroup is self-centeralizing. Journal of Algebra. 2016;**462**:23-36

[19] Dokuchaev M, Exel R, Piccione P, Partial representations and partial group algebras. 1999. arXiv: math/990312

[20] Exel R. Partial actions on groups and actions of inverse semigroups. Proceedings of American Mathematical Society. 1998;**126**:3481-3494

[21] Exel R, Partial Group Actions, Campus de Cantoblanco, Universidad

Aut'onoma de Madrid, Lecture Notes at ICMAT. 2013

[22] Falcon FJ, Nünez RM. Santilli automorhism of partial groups. American Journal of Modern Physics. 2007;**4**(5–1):47-51

[23] Feit W, Thompson J. Solvability of groups of odd order. Pacific Journal of Mathematics. 1963;**13**(3):775-1029

[24] Feldman A. Fitting height of solvable groups admitting fixed point free automorphismgroups. Journal of Algebra. 1978;**53**:268-295

[25] Flavell P. A new proof of the solvable signalizer functor Theorem. Journal of Algebra. 2014;**398**:350-363

[26] Flavell P. Primitive pairs of p-solvable groups. Journal of Algebra. 2010;**324**(4):841-859

[27] Glauberman G. Correspondences of characters for relatively prime operator groups. Candian Journal of Mathematics. 1968;**20**:1465-1488

[28] Glauberman G. On solvable signalizer functors in finite groups. Proceedings of the London Mathematical Society. 1976;**33**(3):1-27

[29] Glauberman G. Prime-power factor groups of finite groups. Mathematische Zeitschrift. 1968;**107**:159-172

[30] Goldschmidt DM. Solvable signalizer functors on finite groups. Journal of Algebra. 1972;**21**:131-148

[31] Goldschmidt DM. 2-signalizer functors on finite groups. Journal of Algebra. 1972;**21**:321-340

[32] Gonzalez A. An extension theory for partial groups and localities. ArXiv: 1507.04392v2. 2015

[33] Gorenstein D. Finite Simple Groups: An Introduction to Their Classification. New York: Plenum Press; 1982

[34] Howie JM. An Introduction to Semigroup Theory. Academic Press; 1976

[35] Humphreys JF. A Course in Group Theory. Oxford: Oxford University Press; 1996

[36] James G, Liebeck M. Representations and Characters of Groups. 2nd ed. Cambridge: Cambridge University Press; 2004

[37] Kurzweil H, Stellmacher B. The Theory of Finite Groups: An Introduction. New York: Universitext, Springer-Verlag; 2004

[38] Lamp C. Duality for partial group action. International Electronic Journal of Algrebra. 2008;**4**:53-62

[39] Martineau R. Elementary abelian fixed point free automorphism groups. The Quarterly Journal of Mathematics. 1972;**23**(2):205-212

[40] Meierfrankenfeld U, Stellmacher B. F-Stability in finite groups. Transactions of the American Mathematical Society. 2009;**361**(5):2509-2525

[41] N'Dao Y, Ayado A. Generalization of Isomorphism Theorems Groups to Partial Groups. HAL archives; 2013 Id: hal-00850152

[42] Shult E. On groups admitting fixed point free operator groups. Illinois Journal of Mathematics. 1956;**9**(4):701-720

[43] Steinberg B. Representation Theory of Finite Groups An Introductory Approach. Universitext. New York: Springer-Verlag; 2012

[44] Suzuki M. Group Theory II. Berlin: Springer-Verlag; 1986

[45] Thompson J. Finite group with fixed point free automorphisms of prime order. Proceedings of the National

Academy of the United States of America. 1959;**45**:578-581

[46] Trudeau R. Introduction to Graph Theory. New York: Kent State University Press; 1996

[47] Vogt RM. Convenient category of topological spaces for algebraic topology. Proceedings of the Advanced Study Institute of Algebra. 1970;**XXII**: 545-555

Chapter 3

Dynamic HUB Selection Process in Wireless Body Area Network (WBAN)

Mahammad Firose Shaik, M. Monica Subhashini and G. Jaya Amrutha

Abstract

Wireless body area network (WBAN), a part of WSN, plays a pivotal role in the remote health monitoring system, these days. Wireless sensor nodes placed in, on, or around the human body are used to create WBAN. This WBAN is mainly used for collecting physiological and vital signals from humans in real-time using sensor nodes. It consists of different sensor nodes and hub, which collects the data from sensor nodes and send them to the gateway. High data rates at HUB cause the damage of an organ receiving high temperature in tissue by electromagnetic signals for a long period. In this chapter, by considering parameters such as the specific absorption rate, Battery Level, Priority of sensor nodes, and signal to noise interference (SINR) a HUB is selected dynamically, which shares the work of the HUB among different sensor nodes. So that workload on HUB decreases and shares its work accordingly to other sensor nodes concerning the data collected through the software LabVIEW. This chapter also illustrates the network (testbed) created using sensors for practically making the change in HUB by using the microcontroller, power, LM 35, BP sensor, Heartrate sensors arranged in a network through Arduino programming. In both these cases, the negative effect of electro-magnetic signals in WBAN, and the tissue damage in humans reduce for remote-health monitoring while increasing the network lifetime.

Keywords: WSN, WBAN, Tissue damage, Dynamic HUB, Specific Absorption Rate, Battery level, Priority, Signal to Inference Ratio, Sensors, LabVIEW, Fuzzy system, Network Life

1. Introduction

Wireless Body Area Network is a part of Wireless sensor network (WSN). Wireless body area network contributes a wide range in health monitoring and broadened its applications considerably [1]. Many lives have been passed down through centuries, due to lack of physical evidence and analyzing the patient up to a tolerance value. WBANs are becoming the limelight in the medical field, expanding their applications in a real-time world by collecting the vitals through implanted sensor nodes. A standard for WBAN as IEEE 802.15.6 is formulated by the IEEE fraternity. This had triggered many types of research to study its uses in WBANs [2].

A tremendous increase in recent technology in the subject of sensing, processing has fueled interest in the technology-based health care system. This is an added advantage to assist an elderly patient [3]. Sensor nodes in WBANs are mounted on the surface or inside the patient skin. These nodes are tiny, light-weighted with the required power and energy. The role of these sensor nodes is to transmit the data from source to destination through the skin as a medium. A loss in the medium occurs where signals or data are absorbed by neighboring tissues, which causes a surge in the temperature of the tissue. Likely, this might cause tissue damage. This is an emerging problem throughout the world, causing the patient to endanger tissue or organ with unbearable pain.

WBAN is lauded for its function through sensor nodes and core nodes, known as HUB which acts as a switchyard between sensor nodes and the gateway. HUB is a fixed sensor node that cannot be varied according to IEEE 802.15.6. The sharing of continuous data through HUB involves higher transmission rates. In contrast, SAR, battery level, Priority of the sensor node, and Signal to Inference Ratio (SINR) are compared to other sensor nodes for better transmission [4].

In this chapter, a HUB is selected dynamically in software based on the selection process through the fuzzy decision systems as shown in **Figure 1**. The fuzzy system converts the numeric value into crisp data where the fuzzy logic is applied and calculates the output based on fuzzy rules. It compares the input data with the standard parametric values of SAR, Battery level, Priority, and SINR (Signal to Inference Ratio). The front panel in LabVIEW is used to collect the data and the processing of fuzzy comparison is done in a block diagram, where every step of data is easy to monitor. The system considers the estimated parameters for different sensor nodes by comparing and gives an estimated sensor node as a HUB.

The effect of changing the HUB dynamically is analytically carried out on a testbed using the microcontroller(ESP-32) which is a 32-bit microcontroller with 34 general purpose IO's, 10 touches sensing IO's, DAC, ADC, pulse counter, UART, I^2C interface, Wi-Fi and Bluetooth as inbuilt application functions, that are mainly used for sending the data(vitals collected) to patients or doctors or neighbors. The sensors used in this testbed are based on application-specific(i.e., in our chapter, we used LM35 for temperature, BMP180 for Blood Pressure measurement, Heartbeat sensor monitoring with finger probe).

The remainder of the work is organized as follows: a description of the relevant work carried out in a focused area is described in the second section. Section 3, gives a glance view of the proposed technique in experimental way and analysis is done. Selection 4, illustrates a detailed overview of the prototype developed,

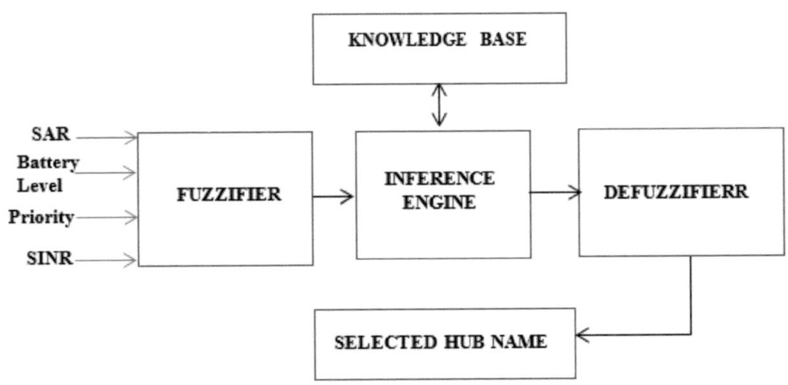

Figure 1.
Fuzzy logic for dynamic HUB selection.

presenting the obtained simulation results and discussion in Section 5. The last section,6, gives a conclusion about the chapter.

2. Related work

The benefits of selecting dynamic HUB through different metrics had a sharp fall down in the damaging of tissues or organs throughout the human body. From the outside, the deployed sensor nodes may be good but practically this causes patients an emergency condition which raises their complexity of the problem.

Not many studies have been made for the reduction of this effect on a sick person. As SAR is considered a major conflict for the rise, studies are made only on estimating and evaluating SAR [5].

By considering the location of the destination node in WBAN as a major part, Ahmed et al. proposed a technique compared to the SAR (Specific Absorption Rate) values under various conditions. It is used to estimate the SAR response on the human body.

Another proposed technique by Wu and Lin to adjust a relay node across the wrist and arm. In return, this will make sensor nodes transmit data packets to HUB with the lowest SAR values and efficient packet rate transmission. This uses an algorithm known as practical swarm optimization to maintain and identify the position of the relay node. No literature survey is added to their works. The performance of HUB is analyzed by Cicioglu and Calhan [6].

Later on, Cicioglu and Calhan made their study on implementing the techniques to maintain a HUB without any loss or damage to survivors [7]. Some other works may be added based on WBAN and SAR issues, but there is no literature study other than a specific absorption-based dynamic HUB selection, This work is extended with the fuzzy-based system in our chapter along with a new additional parameter SINR(Signal to Interference Noise Ratio). This is a major factor while transmitting the data or collecting the information from a patient. If a patient moves or any sensors placed in the network are in motion, then Interference occurs which weakens the network lifetime and manipulates the accurate data. The other major factor causing a signal to be disturbed is noise, which will be eradicated by implementing SINR in our dynamic HUB selection process. When compared to another literature survey, SINR is implemented in this chapter, to reduce the noise, interference and increasing the network lifetime. The proposed method selects a hub based on few parameters along with the signal interference or disturbance during the transmission for better and reliable communication.

2.1 HUB selection process parameters

At first sight, it looked like a typical problem for selecting a hub dynamically in WBANs. Wireless body area network has emerged their development through WSNs. Later on, by estimating the drawbacks of fixed HUB, a need for the selection of dynamic HUB has established. The wireless transmission density is neglected in the evaluating process.

A new hub is selected from the parameters like SAR, Battery Level, Priority of the sensor node, and SINR as in **Table 1**.

The parameter that we seek in the selection process plays an eminent role in the broadcast between sensor nodes and the gateway. This can be sorted for a multipurpose decision-making system using Fuzzy-based rules. The criterions used are SAR, Battery level, Priority, and SINR.

Parameters	Low	Medium	High
SAR (W/kg)	0–1.6	1.6–2.2	2.25–3
BATTERY LEVEL (Joules)	0–1.3	1.4–2	2–5
PRIORITY OF NODES	0–2	2–5	5–8
SINR (dB)	0–12	12–20	20–50

Table 1.
HUB selection parameters table.

If the sensor node has values of SAR less than 1.6 w/kg, battery level in or around the range of 0–4 Joules, the priority of the sensor nodes which is given lowest priority is considered in selecting HUB. And SINR with a range above 20db is best for a new hub. The mentioned parameters above are according to the standard values based on LTE and IEEE fraternity. The illustration of the parameters considered in selecting HUB is depicted below.

2.2 SAR

In today's life, communication devices like mobiles, Wi-Fi have become a common part of the human lifecycle. The electromagnetic interference (EMI) radiation emitted from these devices is mean to be absorbed by a human being at a period of transmitting data. SAR expresses the amount of radiation absorbed by the body, generally represented in terms of Watt per kilogram. The SAR value is differently applicable for children [8]. Any of the communication devices near to tissue can cause damage as it continuously sends data to HUB, so a fuzzy-based rule with membership function as in **Figure 2** is implemented. The specified SAR rate according to FCC is 1.6 W/Kg [5].

This term is generated by

$$\text{Radiation} = \text{d}\left(\frac{dW}{dM}\right)dt \tag{1}$$

Here W is the power (W),
M is the mass (Kg), and.
t is the time (sec).

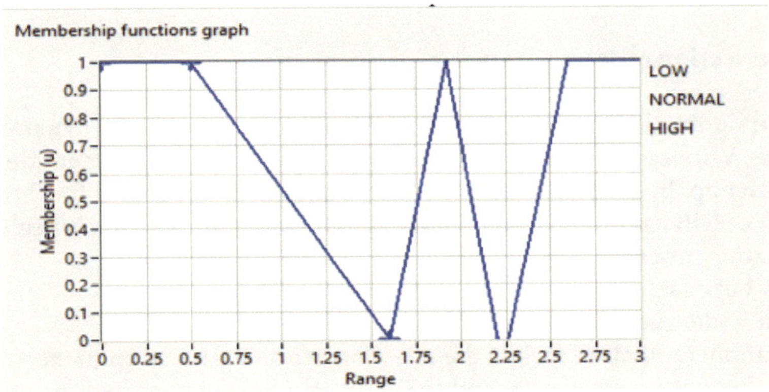

Figure 2.
SAR membership function.

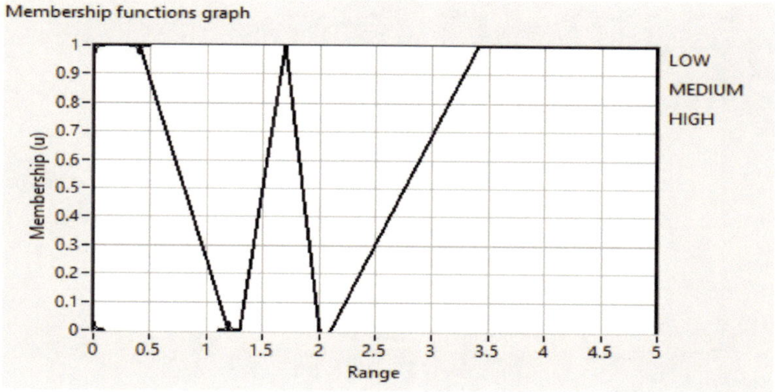

Figure 3.
Battery level range.

2.3 Battery level

The limit used is Battery level for HUB selection is as in **Figure 3**. Sensor nodes Used for WBANs have specific defined energy and power [9].

The sensor nodes can be utilized for a long time if energy is efficiently implemented. As the HUB requires more energy, a fixed hub will lose its required stamina if used for a long time at work. A sensor node is selected as a hub if it is ready to apply the maximum energy after the first sensor node failure, increasing the overall network lifetime.

2.4 Priority of sensor node

A priority of the sensor nodes in WBANs is considered as the parameter in the selection process for the new hub **Figure 4**. The highest priority is given to important vitals like EGG, EEG which transmit data quickly and a need to live for a longer period. In this selection process, the sensor node with the least priority is picked as a hub to transfer high packets.

2.5 SINR

This is the new parameter utilized in our chapter to bring down the interface or disturbances in the network. There are various data traffics in WBAN's, for

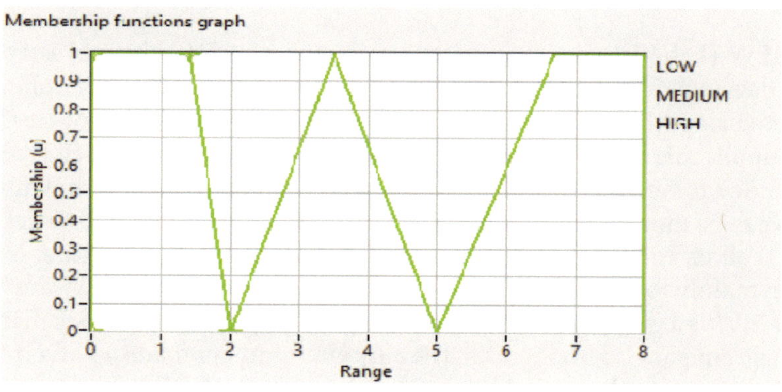

Figure 4.
Priority sensor membership function graph.

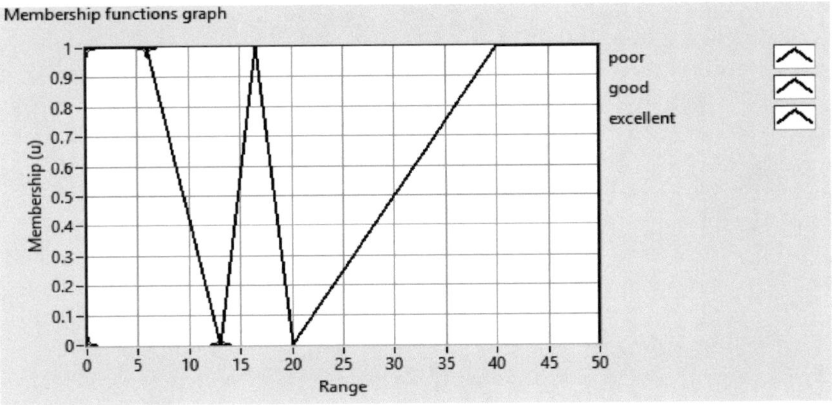

Figure 5.
SINR range in the membership function.

instance, merging of one signal with another or moment of a network from place to place. In wireless communication systems such as networks, SINR is used as the rate of information transfer representing the received signal strength. As the noise in the signal plays an efficient role in decreasing the data accuracy received at the receiver. Moreover, it is defined as the ratio of signal strength to the interference or noise of the signal. For communication link quality, SINR plays a pivotal role. SINR is used for identifying moisture and thermal noise at the sink, in wireless communication if any interference occurs at the data transmission, path loss takes space which results in great network disturbance. So, SINR is utilized to reduce this effect in data transmission.

This is also used to avoid interference in the affected sensor node. Then, a fuzzy-based membership function is used to calculate SINR ranges (0db to 12db, 12db to 20db, 20db to 50db) where the disturbance occurred due to the transmission period is sorted. In Telecommunication LTE measures the values in terms of powers of the signals that is considered as a standard value in wireless transmission of data. The SINR, greater than 20db is selected as HUB as shown in **Figure 5**. This is done by sharply bounded values in fuzzy-based logic for calculating the candidacy value, which decides the HUB using all parameters. The lowest noise signal reflects the highest valued SINR in quality of the network and interference.

3. Experimental results and analysis

LabVIEW (Lab Virtual Instrumentation Engineering Workbench) is a software platform developed by national instruments for data acquisition, controlling, and automation using Microsoft windows, various types of UNIX, Linux Mac OS. It is widely applicable for its features like graphical representation which are highly recommended for engineers and scientists, where data flow can be known and each step can be monitored if required. It also consists of different libraries with different toolkits and the major advantage of this is it is easily configurable with hardware, mainly for processing. The programs operated in LabVIEW are known as Vis. LabVIEW is a geographical programming language to create block diagrams in a block diagram panel. This is a user-friendly environment made up of a front panel and block diagram as shown in **Figure 6** for the dynamic HUB selection process along with the fuzzy-based functions.

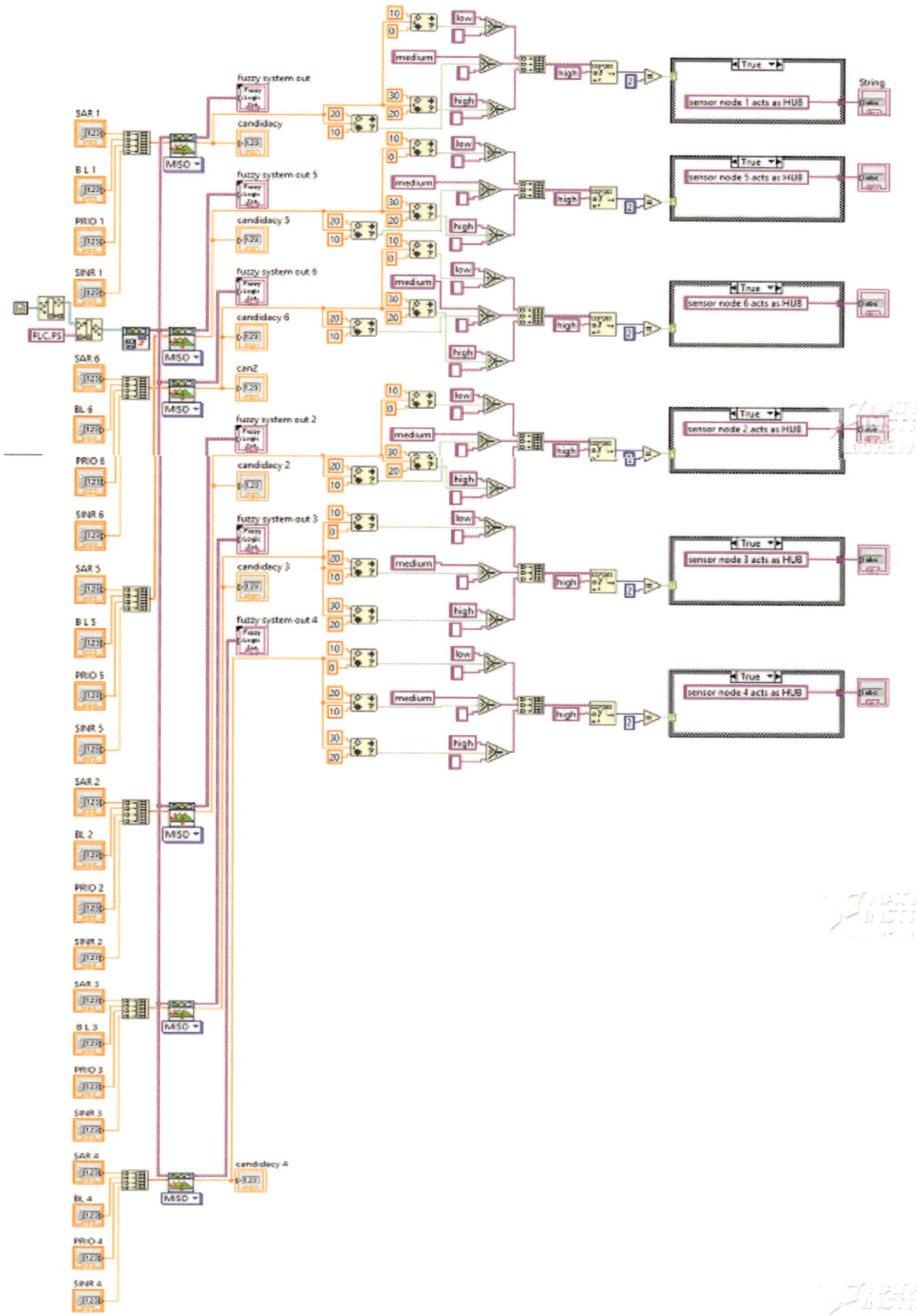

Figure 6.
Block diagram of LabVIEW for dynamic HUB selection.

The below-shown Figure represents the block diagram of the fuzzy-based system(the logic used) for selecting a HUB. Briefly, it collects the sensor data from the parametric values and processes an output by using fuzzy logic. For instance, two sensor nodes are first implemented on the front panel as shown in **Figure 7**.

In the front panel and block diagram, the input is given manually. The block diagram consists of loops and control lines where the flow of data virtually visible.

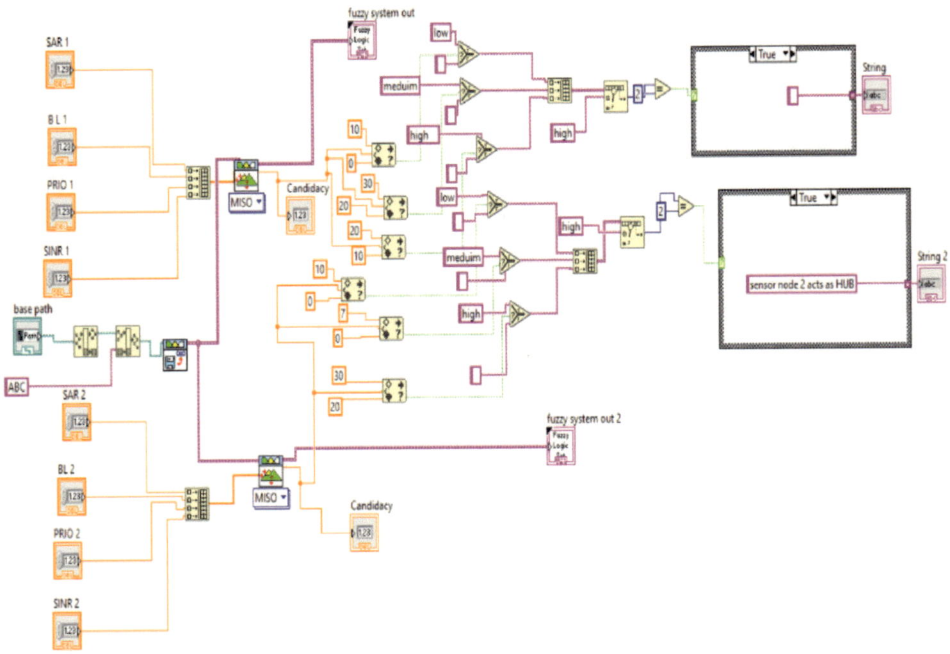

Figure 7.
Front panel code using 2 sensor nodes.

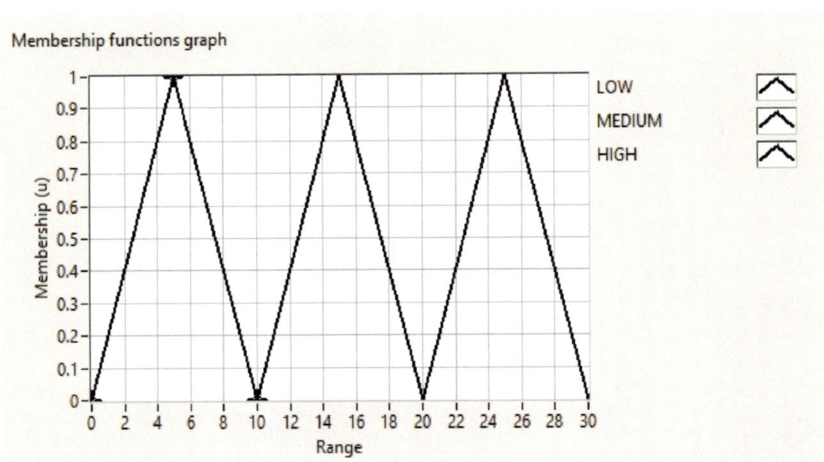

Figure 8.
Candidacy membership function.

The fuzzy system is an inbuilt library function in LabVIEW. The fuzzy system uses an engine to convert numeric data into crisp value as shown in **Figure 1** by implementing the bordered values through fuzzy rules and the output is shown in the candidacy membership function [10].

The Fuzzy output is taken and the common logic is used for a candidacy value with the highest based on the four parameters. The greatest Candidacy value will be considered as a new HUB, Which will be displayed in a string format in the LabVIEW front panel. This output is calculated as in **Figure 8** which considers the low(0–10), medium(10–20), high(20–30) range. Out of which the highest candidacy sensor node will be selected as a Dynamic HUB by using AND or OR functions.

3.1 About fuzzy and its rules

The Fuzzy-logic was initially proposed by a professor, at a university in 1965, named Lofti Zadeh [11]. It is described as a mathematical representation of executed numerical values, would rather say describing words than numbers. In this case, the fuzzy rules set understands the linguistic form of entered parameters i.e., SAR, Battery Level, Priority, SINR, and the controls convert them using an inference engine, membership function, and if-then rules presented in fuzzy rules set. This process is known as Fuzzification. Also, the conversion of if-then rules (which are expandable depending on the user application) to crisp data is known to be a defuzzification process [12]. The candidacy value is formed based on these Fuzzy based rules. There are different formations in which a HUB is selected. Out of the six sensor nodes used, we consider the rules formed are 81 which are depicted in below **Table 2**.

Sl. No.	SINR	SAR	Battery Level	priority	Candidacy level
1	Low	Low	Low	Low	Low
2	Low	Low	Low	Medium	Low
3	Low	Low	Low	High	Low
4	Low	Low	Medium	Low	Low
5	Low	Low	Medium	Medium	Low
6	Low	Low	Medium	High	Low
7	Low	Low	High	Low	Normal
8	Low	Low	High	Medium	Low
9	Low	Low	High	High	Low
10	Low	Medium	Low	Low	Low
11	Low	Medium	Low	Medium	Low
12	Low	Medium	Low	High	Low
13	Low	Medium	Medium	Low	Low
14	Low	Medium	Medium	Medium	Low
15	Low	Medium	Medium	High	Low
16	Low	Medium	High	Low	Low
17	Low	Medium	High	Medium	Low
18	Low	Medium	High	High	Low
19	Low	High	Low	Low	Low
20	Low	High	Low	Medium	Low
21	Low	High	Low	High	Low
22	Low	High	Medium	Low	Low
23	Low	High	Medium	Medium	Low
24	Low	High	Medium	High	Low
25	Low	High	High	Low	Low
26	Low	High	High	Medium	Low
27	Low	High	High	High	Low
28	Medium	Low	Low	Low	Low

Sl. No.	SINR	SAR	Battery Level	priority	Candidacy level
29	Medium	Low	Low	Medium	Low
30	Medium	Low	Low	High	Low
31	Medium	Low	Medium	Low	Low
32	Medium	Low	Medium	Medium	Low
33	Medium	Low	Medium	High	Low
34	Medium	Low	High	Low	High
35	Medium	Low	High	Medium	Low
36	Medium	Low	High	High	Low
37	Medium	Medium	Low	Low	Low
38	Medium	Medium	Low	Medium	Low
39	Medium	Medium	Low	High	Low
40	Medium	Medium	Medium	Low	Low
41	Medium	Medium	Medium	Medium	Low
42	Medium	Medium	Medium	High	Low
43	Medium	Medium	High	Low	Low
44	Medium	Medium	High	Medium	Low
45	Medium	Medium	High	High	Low
46	Medium	High	Low	Low	Low
47	Medium	High	Low	Medium	Low
48	Medium	High	Low	High	Low
49	Medium	High	Medium	Low	Low
50	Medium	High	Medium	Medium	Low
51	Medium	High	Medium	High	Low
52	Medium	High	High	Low	Low
53	Medium	High	High	Medium	Low
54	Medium	High	High	High	Low
55	High	Low	Low	Low	Normal
56	High	Low	Low	Medium	Low
57	High	Low	Low	High	Low
58	High	Low	Medium	Low	Normal
59	High	Low	Medium	Medium	Low
60	High	Low	Medium	High	Low
61	High	Low	High	Low	High
62	High	Low	High	Medium	Normal
63	High	Low	High	High	Normal
64	High	Medium	Low	Low	Low
65	High	Medium	Low	Medium	Low
66	High	Medium	Low	High	Low
67	High	Medium	Medium	Low	Low
68	High	Medium	Medium	Medium	Low
69	High	Medium	Medium	High	Low

Sl. No.	SINR	SAR	Battery Level	priority	Candidacy level
70	High	Medium	High	Low	High
71	High	Medium	High	Medium	Medium
72	High	Medium	High	High	Low
73	High	High	Low	Low	Low
74	High	High	Low	Medium	Low
75	High	High	Low	High	Low
76	High	High	Medium	Low	Low
77	High	High	Medium	Medium	Low
78	High	High	Medium	High	Low
79	High	High	High	Low	Normal
80	High	High	High	Medium	Low
81	High	High	High	High	Low

Table 2.
Fuzzy rules.

4. Prototype for dynamic HUB selection

The Dynamic Hub selection outraced the results of the traditional hub by implementing it in LabVIEW through Fuzzy based system. This is now tested by a prototype developed on the testbed to have a look at the changes in HUB selection concerning the real-time body measurements. As technology has a great impact on health and mobile phones these days become a valuable asset for a normal man. On this point, we connected an app that stores information about the vitals from the human body. Here, an embedded C is evolved to change the HUB dynamically and is provided in a mobile app where the HUB selection is shown in the app. The embedded communication devices used are for the programming of the network as they are designed for a special purpose rather than the general tasks of computers. These are inbuilt with memory storage, a processor, and some peripherals. It can assist using C, ALP, and VB, etc. It also stores the information about the vitals collected for future purposes. As discussed above, the root cause for tissue damage is the temperature rise, working for a longer period in the network. In this prototype, a traditional hub is changed according to the vitals collected from the patient. Based on the functionality and requirement, the embedded system is deployed in our mobiles and so as in this work. The different sensors used in the prototype are described below in **Figure 9**, where the network is placed on the surface of the body.

ESP-32 is a 2.4 GHz microcontroller with both Wi-Fi and Bluetooth combination chip designed for better power and good reliability with a wide range of applications, which consists of an I^2C bus interface, recovery memory, ROM, SRAM, temperature sensor, touch sensor, clock generator, a serial peripheral interface, DAC, ADC, UART (Universal Asynchronous Receiver Transmitter), PWM, Wi-Fi, and Bluetooth peripherals along with radio receiver and transmitter.10 capacitive sensing GPIOs present in the microcontroller are useful for our project as we use a touch sensor for monitoring the heart rate through the illusion of light by finger and the other three sensors are deployed.

The deployment materials used here are a Power supply, Temperature sensor, Heartbeat sensor, Blood pressure monitoring device. The power supply is used to convert high voltage to a low voltage supply, which consists of a transformer,

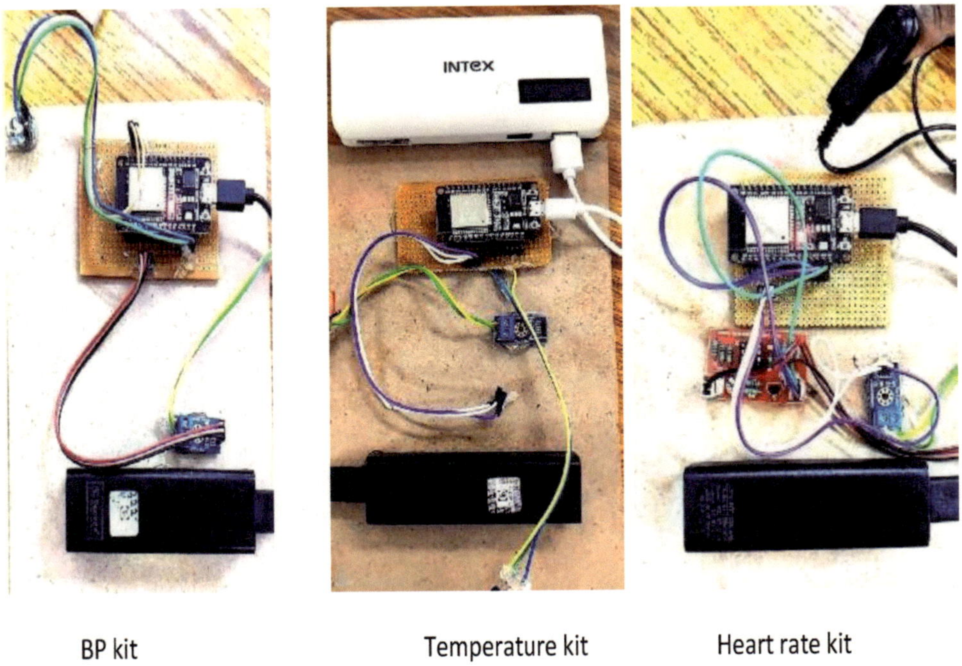

Figure 9.
Hardware kits.

regulator, filter, and rectifier. The temperature sensor, LM 35 is used here to convert the temperature into electrical signals. Yet, it does not require any calibration and best suitable for our work. Having said that, this is considered a low power supply and low self-heat not more than 0.1^0c in still air, most widely used in remote applications. The different male and female connection wires are used for creating a system.

5. Results and discussion

In this section, the results of selecting a HUB in both software and by using a Prototype for decreasing the damage of tissue when placed on the surface of the body are delineated. By using the LabVIEW-based Fuzzy rules, a dynamic HUB is selected by following the standard metric values of parameters. Here, six sensor nodes are considered and shown, which sensor node is changing as a HUB. As LabVIEW is a step-by-step procedure, we can identify the process and rectify if any mistakes happen. Whereas in the prototype developed, we considered three different sensor nodes in a testbed to run the code and examine the HUB selection process dynamically.

5.1 In LabVIEW using fuzzy

In this stage, the drawbacks of using a fixed HUB are rectified by using a dynamic HUB. We considered six sensor nodes for instance and process a dynamic HUB selection using the Fuzzy system in LabVIEW. The Fuzzy-based system consists of if-then rules in the fuzzy decision system.

The parameter values are considered from manual input through the front panel, as it is utilized to display the input and output of any program. The input given starts working in the block diagram through the Fuzzy library function as shown in **Figure 1**. The input of the Fuzzy system, consists of three sub-parts namely, entering the data, membership function, Fuzzy rules, and candidacy output. The candidacy value is sent back to the block diagram through a fuzzy out system with candidacy results and continues the process in a block diagram. As the simulation begins, the sensor node parameter value is considered and sends the information packet from all neighbors to default HUB. The aforementioned parameters are selected based on the universal standard values maintained by their respective standards. The proposed technique using Fuzzy gives an output candidacy value by comparing all sensor nodes' parametric values in the fuzzy decision system. The sensor node with the highest candidacy value I selected as a HUB. The front panel in the software implies the input to a fuzzy engine through a block diagram. Later on, by calculating the graph and candidacy value, a HUB is selected and presented in the front panel in string format.

Here the input is taken through the front panel and the process is done through fuzzy logic presented in block diagram when running the code present in it. If a sensor node has values of SAR less than 1.6 w/kg, battery level in or around the range 0–4 Joules, the SINR greater than 20db, the priority of the sensor nodes is considered and selected as a HUB. The sensor nodes are encountered with different cases such as when all the values are similar or whenever two sensor nodes exhibit the same values, then it is based on the priority of the sensor nodes, which selects the dynamic HUB. The priority of the sensor nodes plays the main role when all other parameters are in balance. In this project, the sensors like ECG, EEG, Temperature are used and temperature being the least sensor as it does not require monitoring continuously, while heartbeat and ECG are given the highest priority whose analysis is required for long hours. This helps in increasing the network lifetime by utilizing the least priority sensors.

The Figure above shown is an example of a front panel result that is selected as a dynamic HUB when placed on the surface of the human body. If any of these values change with time accordingly then the HUB is selected based on the change in parametric values. This makes the fuzzy system easy for multi-systems and increases the overall network lifetime by four times to a fixed hub.

Initially, a random node is selected as the default hub to send the information from source to sink and it is as shown in **Figure 10**. There are certain cases to examine our work. Here, When two sensor nodes of the same values are considered then the fuzzy rules are codded as it predicts the least priority sensor as Dynamic HUB as

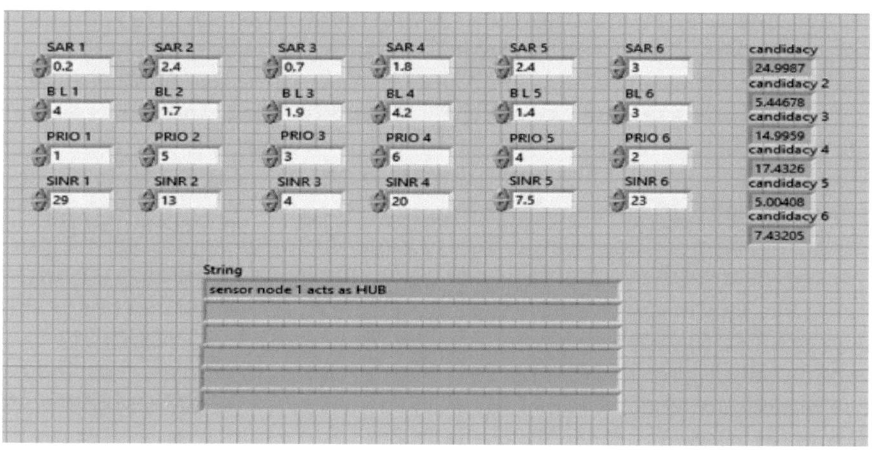

Figure 10.
Default dynamic HUB selected.

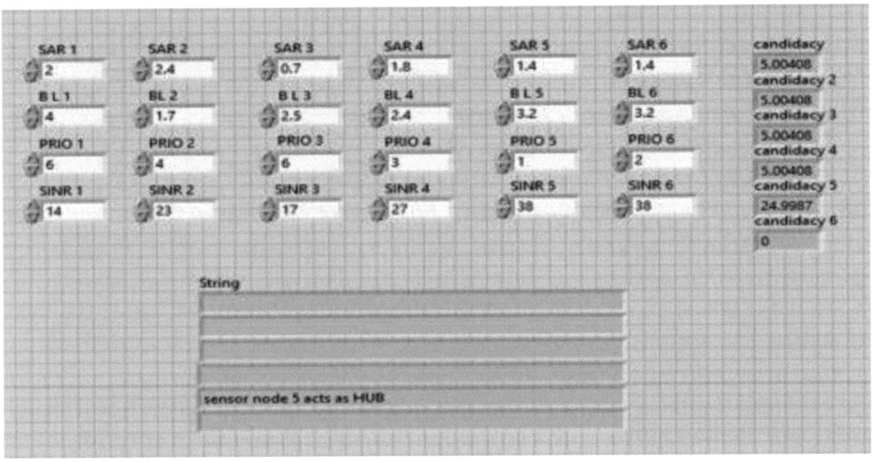

Figure 11.
Similar nodes HUB selection.

said having the least priority leads to un usage of a node for a long time which can enrich the network for the long run. This is laid out in **Figure 11**.

5.2 In prototype using sensors

The hardware kit developed here is made up of three pivotal sensors essential for the human body as of vitals. These are heart rate sensors, Blood pressure, and temperature. The heart rate sensor is a digital output that works using a microcontroller. When a finger is placed, it works on the light modulation on the illusion of blood flow, a red light which transfers from transmitter to receiver, counting the pulses through light flow. Generally used for the measuring of heart rate. Blood Pressure is measured using a mercury column. A Cuff with an oscillatory device in it is wrapped above the upper arm, where it produces vibrations in blood flow in the artery between systole and diastole pressures. The Aurdino Uno on the Arduino Desktop IDE (Arduino software) is used for code. An Arduino board is completely open-source and user-independent which uses a sketchbook, a place to store the programs. The Arduino program is developed by using a delay for our project to identify the sensor node to change as HUB. This can be negotiated for practical usage. The node change is identified by naming the sensors like A, B, and C.

The node change for values change in a mobile app is exposed as in above **Figure 12**. The last sensor used is LM 35, for measuring the temperature of a body. This also uses a microprocessor that converts the input and processed value to digital form for manual purposes. A finger grip for heart rate and a hand cliff BP equipment along with LM 35, temperature sensor are used for deployment of network and are together shown as in **Figure 11**.

The power supply is used for the input source to run the equipment. We can also find, collect or store the vital information collecting in the mobile app for further utility. The change in values for physical movement or activity or exercise results in the change of HUB, which is shown in the mobile app as below in **Figure 13**.

Here, in our chapter Arduino board is used for deploying the program in Micro Processor. A Bluetooth terminal is considered as NODES A, B, and C and performs the program as shown below. In this project, we are supposed to work offline. If we want to use Arduino UNO offline then an Arduino Desktop IDE needs to be installed. The UNO is programmed using the Arduino Software(IDE), our

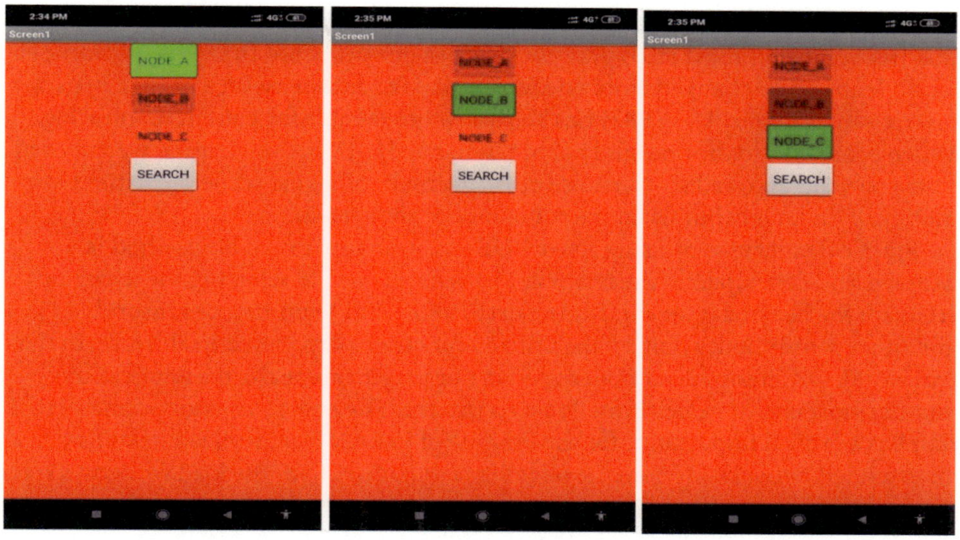

Figure 12.
Dynamic HUB identification.

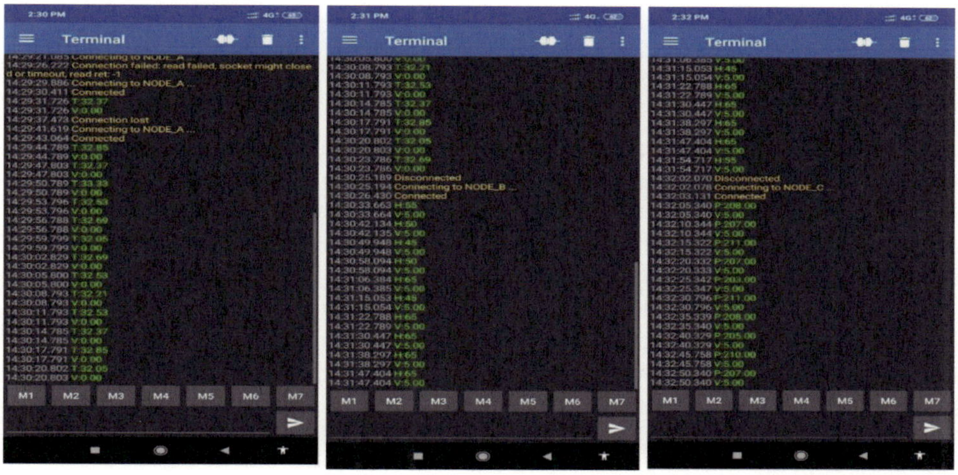

Figure 13.
Information in the node through the mobile app.

integrated Development common to all our boards. Connection is done from UNO board to system is through A B USB cable, sometimes known as a USB printer cable. Furthermore, the sensor nodes are connected with the desired pins in the micro-controller, which runs through the program shown below. The sensors give input to Arduino through a microcontroller, where the code is executed and gives the output in selecting a HUB based on the change in values of collected information. A delay function is used in our program for the identification of HUB change which is not required in real-time health monitoring.

6. Conclusion

The work carried out is to select a dynamic HUB for reducing the damage of tissue when an entire network is placed on a human body wirelessly to collect the data and send it to a receiver through a gateway. This chapter improvises the HUB features regarding fixed HUB drawbacks. With our proposed fuzzy-based framework, Hub can be changed from one sensor node to another with a change in parametric values manually. WBAN has become more reliable for patient monitoring. When compared to the one study about dynamic HUB, in this chapter, we imposed a new parameter to reduce the noise or interference of signal while the patient is in motion. This largely builds on the coherence of the system.

Practically, instead of using fixed HUB traditionally, a variable method is adopted and developed in the testbed. The sensors used in selecting the dynamic HUB are the heartbeat sensor, BP sensor(BP180), and temperature sensor(LM35), and other sensor nodes that can be utilized for other purposes of healthcare, in the real world. Different sensor nodes used in our project are real-time applications and the most useful values in analyzing a patient. The energy consumption of HUB that has a critical importance for the lifetime of WBAN is minimized, resulting in an extended network lifetime. By selecting a dynamic HUB, the load work on a traditional HUB decreases and share the burden among all other sensor nodes, which results the larger network life. To that, the selection of dynamic HUB results is reducing the damage of tissue while increasing the network lifespan. In the future, more parameters can be included either in software and hardware for efficient working of the network.

Acknowledgements

The authors of this Chapter would like to thank each member of Intech-Open resource, the world's leading publisher of open access to books, built by scientists, for scientists. We thank every individual on board for considering our work of Dynamic HUB selection through recent technologies in your "Coding Theory-Recent Advances, New Perspectives, and Applications", by a chapter.

We also like to extend our special thanks to Mr. Josip Knapic, Author Service Manager, who helped us throughout the process by queries and made a way for our financial aid by making a free publication. Without your support, this could not be possible. Thank you everyone for the opportunity to explore our work into a bigger entity. Finally, we would like to thank our friends, family, and colleagues, who supported us consistently over a period for what we are today.

Appendices and nomenclature

Code for selecting a Dynamic HUB.

```
// code for Blood Pressure.
#include"Bluetooth.Serial.h".
#!defined(CONFIG_BT_ENABLED)||!defined(CONFIG_BLUEDROID_
ENABLED).
#ERROR Bluetooth is not enabled! Please run'make menuconfig' to enable it.
#endif.
BluetoothSerail SErialBT;
Void setup() {.
Serail.begin(9600);
SerailBT.begin("NODE_C);//Bluetooth device name.
Serial println("The device started, now you can pair it with Bluetooth!");
}
```

```
Void loop() {.
Float p = analogRead(34);
SerailBT.print("P:" + String(p));
Float v = analogRead(32)/200;
SerailBT.println("v:" + String(v));
Delay(5000);
}
//Code for heart Beat Sensor.
#include"Bluetooth.Serial.h".
#!defined(CONFIG_BT_ENABLED)||!defined(CONFIG_BLUEDROID_
ENABLED).
#ERROR Bluetooth is not enabled! Please run'make menuconfig' to enable it.
#endif.
BluetoothSerail SErialBT;
Int hs = 13;
Int hb = 0;
Void setup() {.
Serial begin(9600);
SerialBT.begin("NODE_B");//Bluetooth device name.
Serial.Println("The device started, now you can pair it with Bluetooth!");
}
Void loop() {.
Double x = millis();
hb = 0;
while(millis() < x + 5000).
{
If(digitalRead(hs)==0).
{
Hb = hb + 1;
Delay(300);
}
}
hb = hb*5;
SerialBT.print("H:" + String(hb));
Float v = analogRead(35)/200;
SerialBT.Println("V:" + String(v));
delay(3000);
}
//Code for Temperature Sensor.
#include"Bluetooth.Serial.h".
#!defined(CONFIG_BT_ENABLED)||!defined(CONFIG_BLUEDROID_
ENABLED).
#ERROR Bluetooth is not enabled! Please run'make menuconfig' to enable it.
#endif.
BluetoothSerail SErialBT;
Void setup() {.
Serial begin(9600);
SerialBT.begin("NODE_A");//Bluetooth device name.
Serial.Println("The device started, now you can pair it with Bluetooth!");
}
Void loop() {.
Float t = (analogRead(34)/2.7)/2.3;
```

```
SerialBT.println("T:" + String(t));
Float v = analogRead(35)/200;
SerialBT.Println("v:" + String(v));
Delay(3000);
}
```


Author details

Mahammad Firose Shaik[1*], M. Monica Subhashini[2] and G. Jaya Amrutha[1]

1 VRSEC, Vijayawada, India

2 School of Electrical Engineering, VIT, Vellore, India

*Address all correspondence to: mahammadfirose.shaik2015@vit.ac.in

References

[1] Latre B.,Braem B.,Moerman I. et al; A Survey on Wireless Body Area Network, Wireless Netws17(1), 1-18, 2011.DOI: 10.1007/s11276-010-0252-4.

[2] Hayeineh T., Almashaqueb G., Ullah S et al: A Survey of wireless technology co-existance in WBAN: analysis and open research Issues. Wireless Network, 20.2165-2199(2014).DOI:10.1007/s11276-014-0736-8.

[3] S. Movassaghi, M. Abolhasan and D. Smith: Cooperative Scheduling with graph coloring for interference mitigation in Wireless Body Area Networks. 2014 IEEE Wireless communications and Networking Confer ence(WCNC),2014,pp.1691-1696, DOI: 10.1109/WCNC.2014.6952484.

[4] Shaik M. F.,Komanapally, V.L .N & Subhashini M, M: A Comparative study of Interference and Mitigation Techniques in WBAN. Wireless Pers Commun 98, 2333-2365,2018, DOI:10.1007/s11277-017-4977-6.

[5] Ahmed A., Halepoto I. A., Khan U. A., Kumar S., and Bhangwar A.R:I-RP: Interference Aware Routing Protocol for WBAN: Mobile Web and Intelligent Information Systems. 2018, Lecture notes in computer science, Vol 10995. Springer,charm. DOI:10.1007/978-3-319-97163-6_6

[6] M.Cicioglu and A. Calhan: Performance analysis of Dynamic HUB Selection Algorithm for WBAN.2016,6 th international conference on control Engineering & Information Technology(CEIT),2018,pp.1-5,DOI: 10.1109/CEIT.2018.8751776

[7] M.Cicioglu and A Calhan: Dynamic HUB selection process Based on Specific Absorption Rate for WBANs.IEEE sensors Journal,vol. 19, no. 14, pp. 5716-5722, 15 july 15, 2019,DOI:10.1109/JSEN.2019.2906044.

[8] R. P. Findlay and P. J. Dimblyow: SAR in children from exposure to wireless local area networks(WLAN).2012. pp. 733-736, DOI: 10.1109/APEMC.2012.6237853

[9] Nikolic, G., Nikolic, T., Stojcev M., Petrovic B., and Jovanovic G: Battery Cpacity Estimation of Wireless Sensor node. In: 2017 IEEE 30th International Conference on Microelectronics(MIEL), pp.279-282

[10] K. P. Adlassing: Fuzzy set theory in medical diagnosis,IEEE Transaction on Systems,Man, Cybernetics, Vol 16, no. 2,pp.260-265,1986, DOI:10.1109/TSMC.1986.4308946.

[11] L. A. Zadeh,: Fuzzy Sets, Fuzzy logic, Fuzzy systems:selected papers by Lofti A. Zadeh, 1996. DOI:10.1142/9789814261302_0021

[12] M. F. Shaik and M. M. Subhashini: Anemia diagnosis by Fuzzy logic using LabVIEW.2017,I2C2.pp.1-5.DOI:10.1109/I2C2.2017.8321790

Chapter 4

How Do Web-Active End-User Programmers Forage?

Sandeep Kaur Kuttal, Abim Sedhain and Benjamin Riethmeier

Abstract

Web-active end-user programmers spend substantial time and cognitive effort seeking information while debugging web mashups, which are platforms for creating web applications by combining data and functionality from two or more different sources. The debugging on these platforms is challenging as end user programmers need to forage within the mashup environment to find bugs and on the web to forage for the solution to those bugs. To understand the foraging behavior of end-user programmers when debugging, we used information forging theory. Information foraging theory helps understand how users forage for information and has been successfully used to understand and model user behavior when foraging through documents, the web, user interfaces, and programming environments. Through the lens of information foraging theory, we analyzed the data from a controlled lab study of eight web-active end-user programmers. The programmers completed two debugging tasks using the Yahoo! Pipes web mashup environment. On analyzing the data, we identified three types of cues: clear, fuzzy, and elusive. Clear cues helped participants to find and fix bugs with ease while fuzzy and elusive cues led to useless foraging. We also identified the strategies used by the participants when finding and fixing bugs. Our results give us a better understanding of the programming behavior of web-active end-users and can inform researchers and professionals how to create better support for the debugging process. Further, this study methodology can be adapted by researchers to understand other aspects of programming such as implementing, reusing, and maintaining code.

Keywords: Information Foraging Theory, End-user programming, Debugging, Visual Programming, Web Mashups

1. Introduction

In modern times, mass communication, mass media, and networking technologies have enabled access to vast amounts of knowledge that are distributed across many continents and time-zones, thus allowing web-active end-users to achieve great feats.

Web-active end-users (also referred to as end-users or end-user programmers) are people who lack programming experience but are engaged in internet activities [1]. There is a substantial number of web-active end-users and their number is continuously growing. The end-users often create applications to complete tasks such as finding apartments to rent in a certain location, tracking flights, and alerting drivers regarding traffic jams. One approach to create such applications is utilizing web mashups programming environments.

Web mashup programming environments allow for creating applications from distributed heterogeneous web sources and functions. Most of the mashup programming environments are visual in nature. Some examples include Yahoo! Pipes [2], IBM mashup maker [3], xfruit [4], Apatar [5], Deri pipes [6], and JackBe [7]. The visual nature of these programming environments allows application creation using code abstraction to ease the programming process. However, the abstraction of code can add complexity of accessing the information, debugging, and comprehending large programs within these environments [1, 8, 9].

Further, end-users create mashup applications by seeking information from the complex ecosystem of the web, which is composed of evolving heterogeneous formats, services, standards, and languages [8]. Seeking information on the web is challenging, as the relevant information is scattered across numerous web sources that end-users must find and manually analyze, an information-seeking problem that costs both time and cognitive effort.

In this chapter, we observe the behavior of end-users while debugging, one of the most difficult aspects of programming [10]. Debugging mashup programs is even more challenging as end-user programmers must locate bugs within the abstract web mashup environment and then locate solutions on the web to fix bugs. The lack of debugging support within mashup environments increases the complexity of finding bugs [9]. Further, finding correct solutions to fix bugs is complicated as the web is a huge compilation of heterogeneous resources.

Currently, it is not clear how web-active end-users seek for bugs in their program and their solutions on the web. Hence, we used an information seeking theory called Information Foraging Theory.

Information Foraging Theory (IFT) can expand our understanding of the information-seeking problems of web-active end-user programmers while debugging. IFT posits that people seek information in the same manner as predators forage for their prey, where predators are the end-users, and the prey is the bugs or bug fixes they are searching for. The hunting grounds or 'patches' where web-active end-users search for these bugs or fixes would be their IDE or the websites they visit and the scents the web-active end-users follow are given by different cues (e.g., links) found on the web [11–15]. IFT has been applied successfully to diverse domains such as documents, the web, user interfaces, and programming environments [15–23].

Past research on web mashups have focused on creating web tools that increase the ease and effectiveness of creating applications by end-user programmers [24–28]. While past IFT research on programming environments has investigated debugging and navigational behavior of professional programmers [19–21]. No prior research exists to understand the debugging behavior of web-active end-user programmers. The only research relevant to this chapter is our own [8], where we created a debugging support for web mashups and investigated the debugging behavior of end-user programmers using IFT with and without the support. Based on this prior research, we found IFT to be the most relevant choice to understand the information-seeking behavior during mashup debugging.

To understand the debugging behavior of end-user programmers we conducted a controlled lab study of eight students who were not computer science majors. The study participants completed their tasks using Yahoo! Pipes, a mashup environment, as it provided the best debugging support at the time. The participants completed two debugging tasks using a think-aloud protocol. We investigated how end-users forage for information within the IDE as well as the web using IFT theory. Our analyses discovered new cues and strategies that end-user programmers pursued while locating the bugs in the mashup environment and foraging the web for fixing the bugs.

This chapter is organized as follows. Section 2 describes the debugging behavior of end-user programmers. Section 3 describes Information Foraging Theory, IFT terminologies from Yahoo! Pipes, and relevant literature. Section 4 describes the background and related work on web mashups, and Yahoo! Pipes. Section 5 describes the methodology and results from the lab study. This section discusses the cues utilized by end-user programmers and their behavior during debugging tasks and provides recommendations. Section 6 summarizes our findings and suggests how web mashup environments can improve the debugging process.

2. Debugging and end-user programmers

Debugging is the process of finding and fixing bugs in the code. Programmers often struggle to debug and hypothesize the "when", "why" and "how" of the bug [29–32]. Debugging is even more challenging for end-user programmers as in one study [33] they spent two-thirds of their time foraging for bugs, while professionals spent only half of their time.

Professionals and end-users use web resources to complete their programming tasks. For example, in one study, novice programmers spent about 19% of their programming time in foraging the web for information such as selecting and using tutorials, searching with synonyms, finding code snippets, and using the web to debug [34], while they spent 35% of their time navigating source code [35]. Vessey [36] investigated both professionals and end-users' debugging approach and found that professionals took a breadth-first approach whereas end-users took a depth-first approach. Our study found that in mashup environments the end user programmers struggle foraging for solutions to bugs on the web.

A major huddle for programmers during debugging is understanding the error messages to fix bugs in the code. Naveed and Sarim [37] analyzed how presentation of error messages affected debugging and programming in IDEs. To fix a bug, first programmers must understand what the error is and where it is located. Mashup environments tend to show errors without much explanation or direction for the end-user to comprehend [9]. End-users struggle to adapt code from tutorials and web forums [38] while fixing bugs. They often struggle with debugging due to lack of knowledge and experience in software engineering and interactive programming environments [39]. Our study confirms that end-user programmers struggle with the lack of or unclear error messages in IDEs.

Understanding end-user programmers' behavior while debugging can help to build better debugging tools that facilitates programming tasks effectively and efficiently. Phalgune et al. [40] studied oracle mistakes - mistakes users make about which values are right and which are wrong - that impact the effectiveness of interactions, testing, and debugging support for end-users. Kuttal et al. [41] added version support to Yahoo! Pipes and investigated how versioning can help end-user programmers to create and debug mashups. Servant et al. [42] create support that allowed panning and zooming of a canvas that contained the snapshots of the code. Myers and Ko suggested various interaction features for IDE to improve debugging such as full visibility of code and timeline visualization of changing values of variables at run-time [43]. Our study helps to understand how end-user programmers debug from a theory perspective that can inform better debugging support for mashup environments.

3. Web mashups

Web mashups allow end-users to build applications by integrating data and functionalities from various web services into a single application. The visual web mashup programming environments facilitate easy creation of applications by end-user programmers who have very little knowledge and experience in programming. Mashup environments provide a full set of functions to the end-users to build new applications.

End-users often create situational mashups as per their specifications [44]. For example, a mashup can take data from Instagram and combine it with Google Maps to display the most recent images and videos of any given location. Users can get the data from APIs, Information Feeds (e.g., Really Simple Syndication (RSS)), or they can collect data by scraping various web pages. Mashup application can be executed within the client's browser, in a server, or combination of both. The advantage of rendering the application in a client's web browser is to give users the opportunity to interact with it. Mashups are popular because of their dynamic content creation and ability to build and share applications through publicly hosted repositories [45].

End-users often develop mashup applications using visual black-box oriented programming environments. Mashup programming environments such as Yahoo! Pipes [2], IBM mashup maker [3], xfruit [4], Apatar [5], Deri Pipes [6], and JackBe [7] provide an easy-to-use visual environment to support the mashup development. Cappiello at el. [46] researched mashup development frameworks oriented towards end-user development to allow users to compose different resources at different levels of granularity relying on the user interface (UI) of the application. Ennals and Gay created MashMaker [24], a tool which allowed end-users to create web mashups without needing to write much code/script. Other mashup creation tools to facilitate end user programmers include MapCruncher [25], Marmite [26], Automator [27], Creo [28], and TreeSheet [47]. Rather than directly studying mashup environments or creating new mashup tools, we qualitatively observe how end-users debug and forage for solutions in programs built in these mashups.

Grammel and Storey [9] investigated various mashup development environments and found lack of debugging support in these environments. Similarly, Stolee and Elbaum [48] studied how we can improve the refactoring of pipe-like mashups, i.e., Yahoo! Pipes for end-users. We focus on understanding end-user programmers' behavior while debugging mashups instead of creating support for mashups.

3.1 Yahoo! Pipes

Now defunct, Yahoo! Pipes was introduced in 2007 and was one of the most popular mashup creation environments that helped users to "rewrite the web" during its existence. During its first year of existence, the Yahoo! Pipes platform executed over 5,000,000 pipes per day. As a visual programming environment, Yahoo! Pipes was well suited for representing the solutions to dataflow-based processing problems. Yahoo! Pipes "programs" helped in combining simple commands together such that the output of one acted as the input for the other. The Yahoo! Pipes engine also facilitated the wiring of modules together and the transfer of data between them.

The Yahoo! Pipes environment was made up of three major components: the canvas, the library (list of modules), and the debugger (refer **Figure 1**). Users used the canvas to create the pipes. The library situated to the left of the canvas, consisted of various modules that were categorized according to functionality. Users dragged modules from the library and placed them on the canvas, then proceeded to connect them to other modules as their need. The debugger, located at the bottom, helped users check the runtime output of the modules.

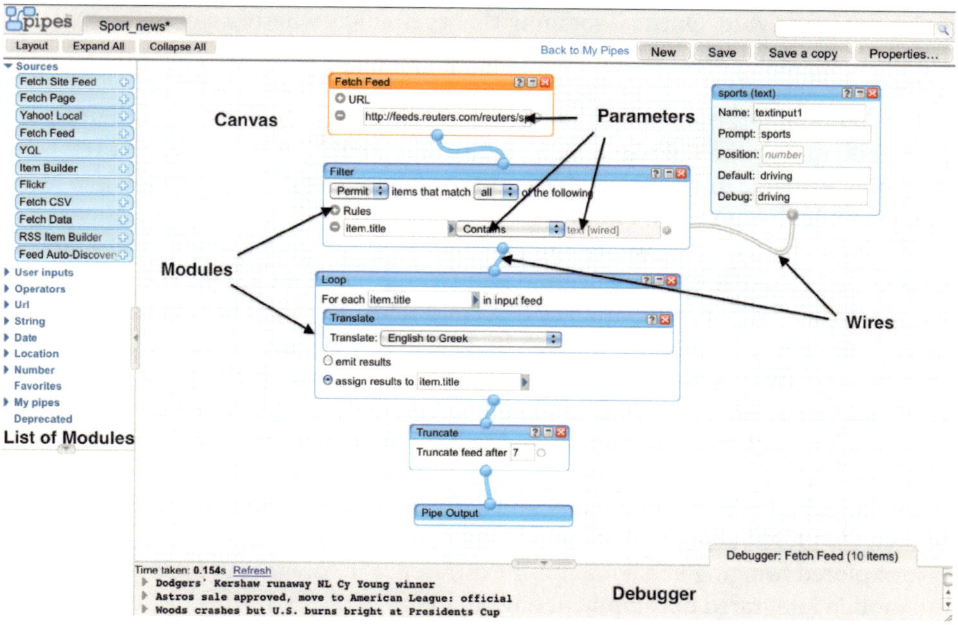

Figure 1.
Yahoo! Pipes.

The inputs and output of the pipes supported different formats. For input, most common formats were APIs, HTML, XML, JSON, RDF, and RSS feeds. Similarly, pipe output formats were RSS, JSON, and KML. The inputs and outputs between modules were primarily RSS feed items consisting of parameters and descriptions. Yahoo! Pipes modules provided manipulation actions that could be executed on these RSS feed parameters. In addition to items, Yahoo! Pipes also allowed datatypes like URL, location, text, number, and date-time to be defined by users.

Figure 1 shows the interface and components of the Yahoo! Pipes environment. The pipe displayed in the figure takes Reuter's Newsfeed (RSS feed) as input using a Fetch Feed module which is then filtered (using a Filter module) based on users' input (sports). These results are converted from English to Greek using a Translate module inside a Loop module. The pipe titles are limited to the first seven results using the truncate module. In **Figure 1**, the debugger window displays the runtime output from the Fetch Feed module.

Yahoo! Pipes allowed the creation and rendering of the pipes on the client side while the executing and storing of the pipe was done on the Yahoo! Servers. The data between the client and server was transfer using JSON format. Yahoo! Pipes allowed end-users to share their pipe (code) as well as reuse other user's pipes by cloning.

Stolee et al. [49] analyzed 32,000 mashups from Yahoo! Pipes repositories based on popularity, configurability, complexity, and diversity. Wang and Wang [50] used Yahoo! Pipes to build a mobile news aggregator application. We used Yahoo! Pipes for this study as it had the best debugging support at the time of the research.

4. Information Foraging Theory and Yahoo! Pipes

Information Foraging Theory (IFT) was developed by Pirolli and Card [11] to understand how people search for information. IFT was inspired by optimal foraging theory, which is a biological theory explaining how predators hunt for their prey in the wild. Optimal foraging theory predicts whether a prey (animal) will try to maximize the energy it gains or minimize the expense to obtain a fixed amount of energy [12]. Similarly, while foraging for information, users must realize their maximum return on information gain at minimum expenditure of their time. Therefore, users, when possible, will modify their strategies to maximize their rate of gaining valuable information [13]. **Table 1** elaborates the IFT terminologies along with examples from Yahoo! Pipes.

IFT has helped to improve the understanding of the users' behaviors and interactions on the web. In the very beginning, research was done for general Internet users, which led to the foundation of IFT [15, 18, 51]. Research has been done to observe and study foragers on the web [8, 15, 21, 51]. IFT has been used to improve the usability of web sites [52] as it has helped to explain and predict why people click a particular link, text, or button on a website [14]. In this research, we qualitatively analyze multiple end-user's foraging behavior to find solutions for their bugs on the web.

IFT has also been used to understand software engineering and software development [8, 19, 20] along with its collaborative environments [17]. Piorkowski et al. have explored foraging behavior and the difference in foraging between desktop and mobile integrated development environment (IDE) [53]. Niu et al. used IFT to design navigation affordances in IDEs [54]. Similarly, IFT has been used to find out the optimal team size for open-source projects [55]. IFT can help to understand the foraging behavior of web-active end-user programmers when engaged in programming activities such as comprehension, reusage of code, implementation, debugging and testing. This research focuses on the debugging behavior of web-active end-user programmers.

Researchers have built computational models of user information foraging behavior when completing tasks [14, 56, 57]. These models have also helped in predicting the effects of social influences on IFT [58]. The researchers have developed

IFT Terminologies	Definitions	Bug Finding (Examples)	Bug Fixing (Examples)
Prey	Bugs; solutions	Finding bug B2 (url does not lead to the right web site) in Fetch Feed module	Finding the correct url and putting it in the Fetch Feed module that contains B2
Information Patch	Localities in the code, documents, examples, web-pages and displays that may contain the prey [23]	Yahoo! Pipes Editor, help documents, help examples	Web pages
Information Feature	Words, links, error messages, or highlighted objects that suggest scent relative to prey	API Key Missing error message "Error fetching [url]. Response: Not found (404)" for bug B1	Finding the right API key from the website
Cues	Proximal links to patches	"about this module" link to the example code related to specific module	"Key" link to the Flickr page to collect the API key
Navigate	Navigation by users through patches	To find bug B2 the user navigated through Yahoo! Pipes editor to external web site	To correct bug B2 participant navigated to various web sites to find the required url

Table 1.
IFT Terminologies from the Yahoo! Pipes Perspective [2].

the WUFIS model for the web [6] and the PFIS model for programmers foraging in IDEs [19, 20]. Ragavan et al. analyzed the novice programmers' foraging in the presence of program variants [22] and built a predictive model [59] inspired by the PFIS model [23, 60]. Our focus is to understand the end-user foraging behavior before creating such computational models.

5. Understanding debugging behavior using an information foraging theory perspective

To understand how end-user programmers forage mashup IDEs (Yahoo! Pipes) for finding bugs and the web for finding solutions for the bugs, we conducted a controlled lab study.

5.1 Lab study using Yahoo! Pipes

Our study observed eight university students who had no background in computer science but had experience with one web language. The students were from diverse fields such as engineering, finance, mathematics, and natural sciences. The participants completed the background questionnaire, a short tutorial on Yahoo! Pipes, and a pilot task to practice programming with Yahoo! Pipes. Once the participants felt comfortable with the Yahoo! Pipes environment, they completed two tasks using the think-aloud method.

The participants were given Yahoo! Pipes programs that were seeded with bugs. The first task (Yahoo! Pipes Error) was a pipe program that was seeded with bugs detected by Yahoo! Pipes and displayed a relevant error message. The second task (Silent Error) was seeded with bugs that were not detected by Yahoo! Pipes and therefore did not display an error message. Further, both tasks contained two classes: top level and nested. Top level contained bugs that were easy to comprehend while the nested class contained sub-pipes with bugs. These sub-pipes needed to be opened in a separate IDE to be found. The details of the tasks can be found in **Table 2**.

Participants' verbalization and actions were transcribed and analyzed using IFT theory. When analyzing the transcripts, we found various cues and strategies used by our participants.

5.2 Types of cues followed by end-user programmers

In finding the bugs and their fixes, participants followed cues. Based on the strength of the cues, they can be classified as clear, fuzzy, and elusive. *Clear cues*

Task	Class	Bugs	Details
Yahoo! Pipes Error	Top Level	B1	API key missing
		B2	Website not found
	Nested	B3	Website not found
Silent Error	Top Level	B4	Website contents changed
		B5	Parameter missing
	Nested	B6	Parameter missing

Table 2.
Details on seeded bugs in the tasks [2].

helped the forager the most as they were easy to understand and provided a direct link to the bugs or their fixes. Hence, they were less costly as they helped participants to spend less time finding and fixing the bugs. *Fuzzy cues* did not have complete information that could lead to a bug. Hence, these cues either lead or mislead to a valuable patch containing prey and were somewhat costly in terms of time spent. *Elusive cues* were very difficult to locate due to absence of direct links to the bugs. These cues were the costliest, as participants often wasted their time foraging for prey in useless patches.

Cues	Description	Example
Clear Cues	Cues that were clear and easy to understand	'*API Key Missing*' cue helped participants look for modules that it was associated with.
Fuzzy Cues	Cues that were difficult to understand	'*org.xml.sax.SAXParseException*' cue was hard for participants to understand as they didn't know what it meant.
Elusive Cues	Cues that were difficult to find	This cue was shown when a fault was nested.

5.3 Debugging behavior of end-user programmers

Participants foraged Yahoo! Pipes IDE to find the bugs and the web to fix the bugs. **Table 3** shows the number of bugs located and fixed by each participant. The results show that end user programmers struggled to debug their pipe programs. The key findings were:

5.3.1 Locating and fixing Yahoo! errors was easier than "silent errors"

The Yahoo! Errors B1 and B2 were easily located by the participants (refer **Table 3**). Yahoo! errors supported clear cues as these bugs had detailed error messages from Yahoo! Pipes. As discussed before, the Yahoo! Pipes environment

Participants	Yahoo! Pipes				Silent Errors							
	B1		B2		B3		B4		B5		B6	
	L	F	L	F	L	F	L	F	L	F	L	F
P1	1	1	—	—	—	—	—	—	1	—	—	—
P2	1	1	1	—	—	—	—	—	—	—	—	—
P3*	1	1	1	1	—	—	1	1	1	1	—	—
P4	1	1	1	—	—	—	1	—	1	1	—	—
P5	1	1	1	—	1	—	1	—	—	—	—	—
P6	1	1	1	—	1	1	1	—	1	—	1	—
P7	1	1	1	—	1	1	1	—	—	—	—	—
P8	1	—	1	—	—	—	1	—	—	—	—	—
Total	8	7	7	1	2	1	6	1	4	2	1	0

**represents a participant with prior knowledge of Yahoo! Pipes.*

Table 3.
Bugs Finding and Fixed per Control Group Participant [2].

provides little support for debugging i.e., just observing the output in the debugger window, hence silent errors B4 and B5 were harder for participants to locate and fix. Hence, end-users' programming IDE should support clear cues i.e., displaying and visualizing of the error messages for the programmers.

5.3.2 Locating bugs was easier than fixing bugs

Locating bugs was easier, especially in the presence of clear cues as well as when participants foraged in the restricted single patch of Yahoo! IDE to locate bugs. But when participants had to fix the bugs, they spent a tremendous amount of time foraging through different web pages (multiple patches). Participants used an enrichment strategy of searching on the web to find the valuable patches. But the quality of their search results depended upon the relevance of keywords. Hence, explicitly stating or automating support of the diet constraints (keywords related to bugs) in the search engines can increase the relevance of the results.

5.3.3 Difficult to locate nested bugs, particularly "silent errors"

The nested bugs were the hardest to locate by the participants as they were elusive. In the case of bug B3, three participants were able to find them as they were *clear cues* with error messages that were returned in the pipe output. To detect the silent errors, participants had to systematically analyze each module of the pipe program and check the debugging window. As a result, only one participant was able to locate the B6 bug. Hence, the IDEs should strengthen the cues by making prey/bugs more visible to the programmers through clear cues.

5.4 Strategies while finding Bugs

Participants foraged for finding the bugs using *Hunting, Enrichment*, and *Navigation* strategies within Yahoo! Pipes IDE.

5.4.1 Hunting strategy

These strategies reflect how the participants hunted for their prey (bugs). The participants had salient goals and they chose cues based on their prominence. For example, they looked for cues in the output of the pipe program. Most participants pursued the first available cue in the output. This explains why most participants pursued bug B1 and B4 (**Table 3**). The participants were mostly unsuccessful in finding the majority of bugs as participants were persistent and pursued a single bug until they found a fault (depth-first search). The hunting strategies were prompted by the environment itself. Hence, designing environments that facilitate problem solving strategies (such as "sleep on the problem") and make prey more visible can facilitate effective hunting strategies by end-user programmers.

5.4.2 Enrichment strategy

To make prey (bugs) more visible as well as to understand the patch, the participants used various enrichment strategies. They realigned/regrouped the modules so that the connections between them were more visible. For exploring the cues, they kept two patches side-by-side. For example, participants placed the editor and documentation side-by-side for better view of each window. This suggests that IDEs should allow multi-context views allowing end user programmers to view different dimensions of code and allow easy manipulation of the environment.

5.4.3 Navigational strategy

The participants carved out regions based on the data flow structure of Yahoo! Pipes and foraged for cues down each path separately. Whenever they found a weak scent (perceived value), they backtracked and returned to the previous cue or patch. Participants often needed to backtrack for small changes, and this suggests supporting fine-grained backtracking that allows non-linear explorations of past programming history [8, 41].

5.5 Strategies followed when fixing bugs

While fixing the bug, participants used *Enrichment, Navigation* and *Verification* strategies.

5.5.1 Enrichment strategy

Participants searched for all possible cues that led them to fixes for the bugs and aggregated them. Most participants used Google to find the solution for bug fixes. They temporarily collected information to reduce cognitive efforts. For example, participants copied original URLs into the notepad and then started making changes to the pipe programs. Hence, supporting to-do lists can help end-user programmers to complete their tasks systematically [61]. Participants also kept the documents (web document and IDE) open side-by-side like when they searched for bugs, necessitating support for multi-contextual views for code and relevant web pages.

5.5.2 Navigational strategy

The participants skimmed through patches for stronger scents. They used already visited patches as negative evidence in their foraging pursuits. For example, participants closed the web pages immediately when they realized they had already visited them. This prompted the participants to backtrack often to previous cues or patches as they were no longer foraging in the right directions. This suggests the need of tools that allow backtracking across multiple patches.

5.5.3 Verification strategy

After fixing the bugs, participants verified it by rerunning the pipe programs and comparing the output to the given solution (oracle). Verification is a very important step in software engineering and building automated techniques to support verification for end-user programmers can help them produce better quality software applications.

6. Conclusions

Our analysis of the debugging behavior of eight end-user participants using information foraging theory suggests that clear cues were the most cost-effective method for finding bugs in mashup environments. Clear cues created stronger perceived value and helped more in the debugging process allowing end-user programmers to locate bugs more easily when compared to fuzzy or elusive cues. Fuzzy and elusive cues resulted in a hindered debugging progress as end-users would end up in useless patches. In addition, the presence of sub-pipes added additional complexity

DOI: http://dx.doi.org/10.5772/intechopen.97765

to the debugging process as participants were unsure where cues were coming from, even if they were clear. Our study also examined how the participants followed the cues to find solutions to the present bugs.

The participants used three main strategies to locate bugs: hunting, navigation, and enrichment. While hunting they used a depth-first strategy resulting in a persistent pursuit of a single bug. When navigating the participants would use the dataflow structure of the program to perceive the value of the bug's location and would backtrack through relevant program histories to locate the bug. Finally, when using the enrichment strategy, participants would organize their environment by placing their IDE side by side with a web browser or by rearranging the code for easier foraging.

The presence of relevant error messages made these strategies for finding bugs more effective; however, when fixing the bugs by foraging the web different strategies were needed in the absence of clear cues. The participants made use of enrichment, navigation, and verification strategies for fixing bugs. They enriched their patches by finding relevant information through Google, storing URLs of useful websites, and by having these resources open side by side next to the editor. The participants navigated the web and used negative evidence to avoid already visited webpages or unhelpful resources. Then by running the program after implementing fixes, the participants would verify that their solutions fixed the bugs.

Our results suggest mashup programming environments need to facilitate clear clues and support hunting, enrichment, navigational, and verification strategies to facilitate the debugging process for end-user programmers.

Author details

Sandeep Kaur Kuttal*, Abim Sedhain and Benjamin Riethmeier
University of Tulsa, Tulsa, OK, USA

*Address all correspondence to: sandeep-kuttal@utulsa.edu

References

[1] Zang N, Rosson MB. What's in a mashup? And why? Studying the perceptions of web-active end users. In: 2008 IEEE Symposium on Visual Languages and Human-Centric Computing 2008 Sep 15 (pp. 31-38). IEEE.

[2] Yahoo! Pipes. [cited 2015May]. Available from: http://pipes.yahoo.com/pipes/

[3] IBM Mashup Maker. [cited 2015May]. Available from: http://www.ibm.com/software/info/mashup-center/

[4] WMaker. [cited 2021Apr8]. Available from: http://www.xfruits.com/

[5] Apatar - Open Source Data Integration & ETL - Apatar - Open Source Data Integration and ETL [Internet]. Apatar Mashup Data Integration. [cited 2021Apr8]. Available from: http://apatar.com/

[6] Deri Pipes. [cited 2015May]. Available from: http://pipes.deri.org/

[7] Jackbe. [cited 2021Apr8]. Available from: https://jackbe.com/

[8] Kuttal SK, Sarma A, Burnett M, Rothermel G, Koeppe I, Shepherd B. How end-user programmers debug visual web-based programs: An information foraging theory perspective. Journal of Computer Languages. 2019 Aug 1; 53y22-37.

[9] Grammel L, Storey MA. A survey of mashup development environments. In: The smart internet 2010 (pp. 137-151). Springer, Berlin, Heidelberg.

[10] Gould JD. Some psychological evidence on how people debug computer programs. International Journal of Man-Machine Studies. 1975 Mar 1;7(2):151-82.

[11] Pirolli P, Card S. Information foraging in information access environments. In: Proceedings of the SIGCHI conference on Human factors in computing systems 1995 May 1 (pp. 51-58)

[12] Kie JG. Optimal foraging and risk of predation: effects on behavior and social structure in ungulates. Journal of Mammalogy. 1999 Dec 6;80(4):1114-29.

[13] Pirolli P, Card S. Information foraging. Psychological review. 1999 Oct;106(4):643.

[14] Chi EH, Pirolli P, Chen K, Pitkow J. Using information scent to model user information needs and actions and the Web. In: Proceedings of the SIGCHI conference on Human factors in computing systems 2001 Mar 1 (pp. 490-497).

[15] Pirolli P., Fu WT. (2003) SNIF-ACT: A Model of Information Foraging on the World Wide Web. In: Brusilovsky P., Corbett A., de Rosis F. (eds) User Modeling 2003. UM 2003. Lecture Notes in Computer Science, vol 2702. Springer, Berlin, Heidelberg.

[16] Burnett MM. Information Foraging Theory in Software Maintenance. OREGON STATE UNIV CORVALLIS; 2012 Sep 30.

[17] Kwan I, Fleming SD, Piorkowski D. Information Foraging Theory for Collaborative Software Development. Corvallis, OR. 2012.

[18] Spool JM, Perfetti C, Brittan D. Designing for the Scent of Information: The Essentials Every Designer Needs to Know About How Users Navigate Through Large Web Sites. User Interface Engineering; 2004.

[19] Lawrance J, Bogart C, Burnett M, Bellamy R, Rector K, Fleming SD. How

programmers debug, revisited: An information foraging theory perspective. IEEE Transactions on Software Engineering. 2010 Dec 23;39(2):197-215.

[20] Lawrance J, Bellamy R, Burnett M. Scents in programs: Does information foraging theory apply to program maintenance?. In: IEEE Symposium on Visual Languages and Human-Centric Computing (VL/HCC 2007) 2007 Sep 23 (pp. 15-22). IEEE.

[21] Jin X, Niu N, Wagner M. Facilitating end-user developers by estimating time cost of foraging a webpage. In2017 IEEE Symposium on Visual Languages and Human-Centric Computing (VL/HCC) 2017 Oct 11 (pp. 31-35). IEEE

[22] Srinivasa Ragavan S, Kuttal SK, Hill C, Sarma A, Piorkowski D, Burnett M. Foraging among an overabundance of similar variants. In: Proceedings of the 2016 CHI Conference on Human Factors in Computing Systems 2016 May 7 (pp. 3509-3521).

[23] Lawrance J, Bellamy R, Burnett M, Rector K. Using information scent to model the dynamic foraging behavior of programmers in maintenance tasks. In: Proceedings of the SIGCHI Conference on Human Factors in Computing Systems 2008 Apr 6 (pp. 1323-1332).

[24] Ennals R, Gay D. User-friendly functional programming for web mashups. In: Proceedings of the 12th ACM SIGPLAN international conference on Functional programming 2007 Oct 1 (pp. 223-234).

[25] Elson J, Howell J, Douceur JR. MapCruncher: integrating the world's geographic information. ACM SIGOPS Operating Systems Review. 2007 Apr 1;41(2):50-9.

[26] Wong J, Hong J. Marmite: end-user programming for the web. InCHI'06 extended abstracts on Human factors in computing systems 2006 Apr 21 (pp. 1541-1546).

[27] Automator User Guide for Mac [Internet]. Apple Support. [cited 2021Apr8]. Available from: https://support.apple.com/guide/automator/welcome/mac

[28] Faaborg A, Lieberman H. A goal-oriented web browser. In: Proceedings of the SIGCHI conference on Human Factors in computing systems 2006 Apr 22 (pp. 751-760).

[29] LaToza TD, Myers BA. Developers ask reachability questions. In: Proceedings of the 32nd ACM/IEEE International Conference on Software Engineering-Volume 1 2010 May 1 (pp. 185-194).

[30] Fitzgerald S, McCauley R, Hanks B, Murphy L, Simon B, Zander C. Debugging from the student perspective. IEEE Transactions on Education. 2009 Sep 15;53(3):390-6.

[31] Ko AJ, Myers BA. Finding causes of program output with the Java Whyline. In: Proceedings of the SIGCHI Conference on Human Factors in Computing Systems 2009 Apr 4 (pp. 1569-1578).

[32] Ko AJ, Myers BA. Designing the whyline: a debugging interface for asking questions about program behavior. In: Proceedings of the SIGCHI conference on Human factors in computing systems 2004 Apr 25 (pp. 151-158).

[33] Cao J, Rector K, Park TH, Fleming SD, Burnett M, Wiedenbeck S. A debugging perspective on end-user mashup programming. In2010 IEEE Symposium on Visual Languages and Human-Centric Computing 2010 Sep 21 (pp. 149-156). IEEE.

[34] Brandt J, Guo PJ, Lewenstein J, Dontcheva M, Klemmer SR. Two studies

of opportunistic programming: interleaving web foraging, learning, and writing code. In: Proceedings of the SIGCHI Conference on Human Factors in Computing Systems 2009 Apr 4 (pp. 1589-1598).

[35] Ko AJ, Myers BA, Coblenz MJ, Aung HH. An exploratory study of how developers seek, relate, and collect relevant information during software maintenance tasks. IEEE Transactions on software engineering. 2006 Nov 30;32(12):971-87.

[36] Vessey I. Expertise in debugging computer programs: A process analysis. International Journal of Man-Machine Studies. 1985 Nov 1;23(5):459-94.

[37] Naveed MS, Sarim M. Analyzing the Effects of Error Messages Presentation on Debugging and Programming. Sukkur IBA Journal of Computing and Mathematical Sciences. 2021 Jan 5;4(2):38-48.

[38] Brandt J, Guo PJ, Lewenstein J, Dontcheva M, Klemmer SR. Two studies of opportunistic programming: interleaving web foraging, learning, and writing code. In: Proceedings of the SIGCHI Conference on Human Factors in Computing Systems 2009 Apr 4 (pp. 1589-1598).

[39] Ruthruff JR, Burnett M. Six challenges in supporting end-user debugging. ACM SIGSOFT Software Engineering Notes. 2005 May 21;30(4):1-6.

[40] Phalgune A, Kissinger C, Burnett M, Cook C, Beckwith L, Ruthruff JR. Garbage in, garbage out? An empirical look at oracle mistakes by end-user programmers. In: 2005 IEEE Symposium on Visual Languages and Human-Centric Computing (VL/HCC'05) 2005 Sep 20 (pp. 45-52). IEEE.

[41] Kuttal SK, Sarma A, Rothermel G. On the benefits of providing versioning support for end users: an empirical study. ACM Transactions on Computer-Human Interaction (TOCHI). 2014 Feb 1;21(2):1-43.

[42] Servant F. Supporting bug investigation using history analysis. In: 2013 28th IEEE/ACM International Conference on Automated Software Engineering (ASE) 2013 Nov 11 (pp. 754-757). IEEE.

[43] Myers B, Ko A. Studying development and debugging to help create a better programming environment. In: CHI 2003 Workshop on Perspectives in End User Development 2003 Apr (pp. 65-68). FL: Fort Lauderdale.

[44] Jones MC, Churchill EF. Conversations in developer communities: a preliminary analysis of the yahoo! pipes community. In: Proceedings of the fourth international conference on Communities and technologies 2009 Jun 25 (pp. 195-204).

[45] Huang AF, Huang SB, Lee EY, Yang SJ. Improving end-user programming with situational mashups in web 2.0 environment. In2008 IEEE International Symposium on Service-Oriented System Engineering 2008 Dec 18 (pp. 62-67). IEEE.

[46] Cappiello C, Matera M, Picozzi M. A UI-centric approach for the end-user development of multidevice mashups. ACM Transactions on the Web (TWEB). 2015 Jun 16;9(3):1-40.

[47] Leonard TA. Tree-sheets and structured documents (Doctoral dissertation, University of Southampton).

[48] Stolee KT, Elbaum S. Refactoring pipe-like mashups for end-user programmers. In: Proceedings of the 33rd International Conference on Software Engineering 2011 May 21 (pp. 81-90).

[49] Stolee KT, Elbaum S, Sarma A. Discovering how end-user programmers and their communities use public repositories: A study on Yahoo! Pipes. Information and Software Technology. 2013 Jul 1;55(7):1289-303.

[50] Wang HB, Wang ZH. Building Mobile News Aggregation Application with Yahoo Pipes. In: Advanced Materials Research 2013 (Vol. 756, pp. 1943-1947). Trans Tech Publications Ltd.

[51] Card SK, Pirolli P, Van Der Wege M, Morrison JB, Reeder RW, Schraedley PK, Boshart J. Information scent as a driver of web behavior graphs: Results of a protocol analysis method for web usability. In: Proceedings of the SIGCHI conference on Human factors in computing systems 2001 Mar 1 (pp. 498-505).

[52] Chi EH, Rosien A, Supattanasiri G, Williams A, Royer C, Chow C, Robles E, Dalal B, Chen J, Cousins S. The bloodhound project: automating discovery of web usability issues using the InfoScentπ simulator. In: Proceedings of the SIGCHI conference on Human factors in computing systems 2003 Apr 5 (pp. 505-512).

[53] Piorkowski D, Penney S, Henley AZ, Pistoia M, Burnett M, Tripp O, Ferrara P. Foraging goes mobile: Foraging while debugging on mobile devices. In2017 IEEE Symposium on Visual Languages and Human-Centric Computing (VL/HCC) 2017 Oct 11 (pp. 9-17). IEEE.

[54] Niu N, Mahmoud A, Bradshaw G. Information foraging as a foundation for code navigation (NIER track). In: Proceedings of the 33rd International Conference on Software Engineering 2011 May 21 (pp. 816-819).

[55] Bhowmik T, Niu N, Wang W, Cheng JR, Li L, Cao X. Optimal group size for software change tasks: A social information foraging perspective. IEEE transactions on cybernetics. 2015 Apr 22;46(8):1784-95.

[56] Fu WT, Pirolli P. SNIF-ACT: A cognitive model of user navigation on the World Wide Web. Human–Computer Interaction. 2007 Nov 1;22(4):355-412.

[57] Chi EH, Pirolli P, Pitkow J. The scent of a site: A system for analyzing and predicting information scent, usage, and usability of a web site. In: Proceedings of the SIGCHI conference on Human factors in computing systems 2000 Apr 1 (pp. 161-168).

[58] Pirolli P. Information foraging theory: Adaptive interaction with information. Oxford University Press; 2007 Apr 12.

[59] Ragavan SS, Pandya B, Piorkowski D, Hill C, Kuttal SK, Sarma A, Burnett M. PFIS-V: modeling foraging behavior in the presence of variants. In: Proceedings of the 2017 CHI Conference on Human Factors in Computing Systems 2017 May 2 (pp. 6232-6244).

[60] Lawrance J, Burnett M, Bellamy R, Bogart C, Swart C. Reactive information foraging for evolving goals. In: Proceedings of the SIGCHI Conference on Human Factors in Computing Systems 2010 Apr 10 (pp. 25-34).

[61] Grigoreanu VI, Burnett MM, Robertson GG. A strategy-centric approach to the design of end-user debugging tools. In: Proceedings of the SIGCHI Conference on Human Factors in Computing Systems 2010 Apr 10 (pp. 713-722).

Chapter 5

A Public Key Cryptosystem Using Cyclotomic Matrices

Md. Helal Ahmed, Jagmohan Tanti and Sumant Pushp

Abstract

Confidentiality and Integrity are two paramount objectives in the evaluation of information and communication technology. In this chapter, we propose an arithmetic approach for designing asymmetric key cryptography. Our method is based on the formulation of cyclotomic matrices correspond to a diophantine system. The strategy uses in cyclotomic matrices to design a one-way function. The result of a one-way function that is efficient to compute, however, is hard to process its inverse except if privileged information about the hidden entry is known. Also, we demonstrate that encryption and decryption can be efficiently performed with the asymptotic complexity of $\mathcal{O}(e^{2.373})$. Finally, we study the computational complexity of the cryptosystem.

Keywords: finite fields, discrete logarithm problem, cyclotomic numbers, cyclotomic matrix, public key, secret key

1. Introduction

Apart from a rich history of Message encryption, the cryptosystem became more popular in the twentieth century upon the evolution of information technology. Until the last part of the 1970s, all cryptographic message was sent by the symmetric key. This implies somebody who has sufficient data to encode messages likewise has enough data to decode messages. Consequently, the clients of the framework must have to impart the secret key furtively. As a result of an issue stealthily key sharing, Diffie and Hellman [1] developed a totally new sort of cryptosystem called public key cryptosystem.

In a Public key cryptosystem, both parties (in a two-party system) have a pair of public enciphering and secret deciphering keys [2, 3]. Any party can send encrypted messages to an assigned party using a public enciphering key. However, only the assigned party can decrypt the message utilizing their corresponding secret deciphering key [4]. After that various public key cryptosystems were introduced based on tricky mathematical problems. Among these, RSA is the longest reasonable use of cryptography. Since its design, in spite of all effort, it has not been broken yet. The security of the RSA is acknowledged to be established on the issue of the factorization of an enormous composite number. Be that as it may, there are some practical issues in RSA execution. The fundamental issue is the key arrangement time that is absurdly long for computationally restricted processors used in certain applications. Another issue is the size of the key. It was demonstrated that the time required to factor an n-bit integer by *index calculus factorization* technique is of order $2^{n^{1/2+\delta}}, \delta > 0$ [5]. In 1990's, J. Pollard [6] demonstrated that it was possible in time bounded by $2^{n^{1/3+\delta}}, \delta > 0$. The reduction of the exponent of n has significant outcomes over the long run. It should likewise be expanded each year as a result of upgrades in the factorization calculations and computational power. Until 2015, it was prescribed the base size of the RSA key should be 1024 bits and subsequently increases to 4096 & 8192 bits by 2015 & 2025 respectively [7]. While trying to remedy these issues, Discrete logarithm problem (DLP) has been utilized (to reduce key setup time and size of the key).

Discrete logarithm problem (DLP) is a mathematical problem that occurs in many settings and it is hard to compute exponent in a known multiplicative group [8]. Diffie-Hellman [1], ElGamal [9], Digital Signature Algorithm [10], Elliptic curve cryptosystems [11, 12] are the schemes evolved under the Discrete logarithm algorithm. The security of Diffie-Hellman relied upon the complexity of solving the discrete logarithm problem. However, the scheme has some disadvantages. It has not been demonstrated that breaking the Diffie-Hellman key exchange has relied upon DLP and also the scheme is vulnerable to a man-in-the-middle attack. For the security aspect, cryptosystem [9] was proposed, to introduce a digital signature algorithm (DSA) that's primarily based on Diffie-Hellman DLP and key distribution scheme. It was demonstrated that DSA is around multiple times littler than the RSA signature and later DSA has been supplanted by the elliptic curves digital signature algorithm (ECDSA). Nonetheless, it has some practical implementation problems [13–15]. The length of the smallest signature is of 320 bits, which is still being too long for computationally restricted processors. Another issue emerged is as a correlation with RSA in a field with prime characteristics, which is forty times slower than RSA [16].

There are some other designs for public-key cryptosystems based on some extensive features of matrices. However, there were some practical implementation problems. Thus it had never achieved wide popularity in the cryptographic community. McElice [17] come up with a public key cryptosystem rooted on the Goppa codes Hamming metric. The scheme has the advantage that it has two to three orders of magnitude faster than RSA. Despite its advantage, it has some drawbacks. It was demonstrated that the length of the public key is 2^{19} bits and the data expansion is too large. Some other extensions of the scheme can also be found in [18–20]. Unfortunately, the scheme & its variants has been broken in [21–23]. Later, Gabidulin [24] come up with the rank metric & the Gabidulin codes over a finite field with q elements, where $q = p^r$ i.e. $\mathbf{F}_q$, as an alternative for the Hamming metric. The efficiency of the scheme relied on same set of parameters and the complexity of the decoding algorithm for random codes in rank metric is tons higher than the Hamming metric [17, 25–27]. Numerous fruitful attacks were utilized on the structure of the public code [28–30]. To prevent these attacks, numerous alterations of the cryptosystems were made, consequently drastically increases the size of the key [31–33]. Lau and Tan [34] proposed new encryption with a public key matrix by considering the addition of a random distortion matrix over $\mathbf{F}_q$ of full column rank n. There are also many other design on matrices, which are not cited here, but none of them gain wide popularity in the cryptographic community due to lack of efficient implementation problems in one and another way.

Thinking about these inadequacies, it would be desirable to have a cryptosystem dependent on other than the presumptions as of now being used. Thus, we propose a cyclotomy asymmetric cryptosystem (CAC) based on strong assumptions of DLP that have to reduce the key size and faster the computational process.

1.1 Outline of our scheme

In this chapter, we consider two significant problems in the theory of cyclotomic numbers over $\mathbf{F}_p$. The first one deals with an efficient algorithm for fast computation of all the cyclotomic numbers of order $2l^2$, where l is prime. The subsequent one deals with designing public key cryptosystem based on cyclotomic matrices of order $2l^2$. The strategy employs for designing public-key cryptosystem utilizing cyclotomic matrices of order $2l^2$, whose entries are cyclotomic numbers of order $2l^2$, l be prime, where cyclotomic numbers are certain pairs of solutions $(a, b)_{2l^2}$ of order $2l^2$ over a finite field $\mathbf{F}_p$ with p elements.

In our approach, to designing cyclotomy asymmetric cryptosystem (CAC) based on trapdoor one-way function (OWF). The public key is obtained by choosing a non-trivial generator $\gamma \in \mathbf{F}_p^*$. The chosen value of the generator constructs a cyclotomic matrix of order $2l^2$. It is believed that cyclotomic matrices of order $2l^2$ is always non-singular if the value of $k > 1$. Since there are efficient algorithms for the construction of cyclotomic matrices. Consequently, the key setup time in our proposed cryptosystem is much shorter than previously designed cryptosystems.

In the scheme, the secret key is given by choosing a different non-trivial generator, which is accomplished by discrete logarithm problem (DLP) over a finite field $\mathbf{F}_p^*$. A key-expansion algorithm is employed to expand the secret keys, which form a non-singular matrix of order $2l^2$. Here it is important to note that, if one can change the generators of $\mathbf{F}_p^*$, then entries of cyclotomic matrices get interchanged among themselves, however, the nature of the cyclotomic matrices remain as same. The decryption algorithm involves efficient algebraic operations of matrices. Hence the decryption in our proposed CAC is very efficient. In view of the perspective on the efficient encryption and decryption features, the polynomial time algorithm ensures that the proposed CAC makes it attractive in computationally restricted processors.

The chapter is organized as follows: Section 2 presents the definition and notations, including some well-known properties of cyclotomic numbers of order $2l^2$. Section 3 presents the construction of cyclotomic matrices of order $2l^2$. Section 4 contains encryption and decryption algorithms of CAC along with a numerical example. In addition, the computational complexity of the proposed CAC is discussed and in Section 5 presents the encryption & decryption can be efficiently perform with asymptotic complexity of $\mathcal{O}(e^{2.373})$. Finally, a brief conclusion is reflected in Section 6.

2. Cyclotomic numbers

Cyclotomic numbers are one of the most vital objects in Number Theory. These numbers had been substantially utilized in Cryptography, Coding Theory and other branches of Information Theory. Thus, calculation of cyclotomic numbers, so called to as cyclotomic number problems, of various orders is one of the primary problems in Number Theory. Complete answers for cyclotomic number problem for $e = 2 - 6$, 7, 8, 9, 10, 11, 12, 14, 15, 16, 18, 20, 22, l, $2l$, l^2, $2l^2$ with l an odd prime had been investigated by many authors see ([35–40] and the references there in). The section contains the generalized definition of cyclotomic numbers of order e, useful notations followed by properties of cyclotomic numbers of order $2l^2$. These properties play a vital role in determining which cyclotomic numbers of order $2l^2$ are sufficient for the determination of all $4l^4$ cyclotomic numbers of order $2l^2$. The section also examines the cyclotomic matrices of order $2l^2$.

2.1 Definition and notations

Let $e \geq 2$ be an integer, and $p \equiv 1(\text{mod} e)$ an odd prime. One writes $p = ek + 1$ for some positive integer k. Let $\mathbf{F}_p$ be the finite field of p elements and let γ be a generator of the cyclic group $\mathbf{F}_p^*$. For $0 \leq a, b \leq e - 1$, the cyclotomic number $(a, b)_e$ of order e is defined as the number of solutions (s, t) of the following:

$$\gamma^{es+a} + \gamma^{et+b} + 1 \equiv 0 \ (\text{mod}\, p); \qquad 0 \leq s, t \leq k - 1. \tag{1}$$

2.2 Properties of cyclotomic numbers of order $2l^2$

In this subsection, we recalled some elementary properties of cyclotomic numbers of order $2l^2$ [38]. Let $p \equiv 1 \ (\text{mod} 2l^2)$ be a prime for an odd prime l and we write $p = 2l^2k + 1$ for some positive integer k. It is clear that $(a, b)_{2l^2} = (a', b')_{2l^2}$ whenever $a \equiv a' \ (\text{mod} 2l^2)$ and $b \equiv b' \ (\text{mod} 2l^2)$ as well as $(a, b)_{2l^2} =$ $(2l^2 - a, b - a)_{2l^2}$. These imply the following:

$$(a, b)_{2l^2} = \begin{cases} (b, a)_{2l^2} & \text{if } k \text{ is even,} \\ (b + l^2, a + l^2)_{2l^2} & \text{if } k \text{ is odd.} \end{cases} \tag{2}$$

Applying these facts, one can check that

$$\sum_{a=0}^{2l^2-1} \sum_{b=0}^{2l^2-1} (a, b)_{2l^2} = q - 2 \tag{3}$$

and

$$\sum_{b=0}^{2l^2-1} (a, b)_{2l^2} = k - n_a, \tag{4}$$

where n_a is given by

$$n_a = \begin{cases} 1 & \text{if } a = 0, \ 2 \mid k \text{ or if } a = l^2, \ 2 \nmid k, \\ 0 & \text{otherwise.} \end{cases}$$

3. Cyclotomic matrices

This section presents the procedure to determine cyclotomic matrices of order $2l^2$ for prime l. We determine the equality relation of cyclotomic numbers and discuss how few of the cyclotomic numbers are enough for the construction of whole cyclotomic matrix. Further generators for a chosen value of p will be determined followed by the generation of a cyclotomic matrix. At every step, we have included a numerical example for the convenience to understand the procedure easily.

Definition:- Cyclotomic matrix of order $2l^2$, l be a prime, is a square matrix of order $2l^2$, whose entries are pair of solutions $(a,b)_{2l^2}$; $0 \le a,b \le 2l^2 - 1$, of the Eq. (1).

For instance **Table 1** depicts a typical cyclotomic matrix of order 8 (assuming $l = 2$). Whose construction steps have been given in the next subsection.

3.1 Construction of cyclotomic matrix

Typically construction of a cyclotomic matrix has been subdivided into four subsequent steps. Below are those ordered steps for the construction of a cyclotomic matrix;

1. For given l, choose a prime p such that p satisfies $p = 2l^2k + 1$, $k \in \mathbf{Z}^+$. The initial entries of the cyclotomic matrix are the arrangement of pair of numbers $(a,b)_{2l^2}$ where a and b usually vary from 0 to $2l^2 - 1$.

2. Determine the equality relation of pair of $(a,b)_{2l^2}$, which reduces the complexity of pair of solution $(a,b)_{2l^2}$ of Eq. (1), that is discuss in next sub-section.

3. Determine the generators of chosen p (i.e. generators of $\mathbf{F}_p^*$). Let $\gamma_1, \gamma_2, \gamma_3, \dots,$ γ_n be generators of $\mathbf{F}_p^*$.

4. Choose a generator (say γ_1) of $\mathbf{F}_p^*$ and put in Eq. (1). This will give cyclotomic matrix of order $2l^2$ w.r.t. chosen generator γ_1.

The first step initializes the entries of cyclotomic matrix of order $2l^2$. Value of p will be determined for given l. Assuming $l = 2$, an example of such initialization of matrix of order 8 has been shown in **Table 1**.

For the construction of cyclotomic matrix, it does not require to determine all the cyclotomic numbers of a cyclotomic matrix which is shown in **Table 1** [36]. By well-known properties of cyclotomic numbers of order $2l^2$, cyclotomic numbers are divided into various classes, therefore there are a pair of the relation between the entries of initial table (**Table 1**) of a cyclotomic matrix. Thus to avoid calculating the same solutions in multiple times, we determine the equality relation of

(a,b)				**b**				
a	**0**	**1**	**2**	**3**	**4**	**5**	**6**	**7**
0	(0,0)	(0,1)	(0,2)	(0,3)	(0,4)	(0,5)	(0,6)	(0,7)
1	(1,0)	(1,1)	(1,2)	(1,3)	(1,4)	(1,5)	(1,6)	(1,7)
2	(2,0)	(2,1)	(2,2)	(2,3)	(2,4)	(2,5)	(2,6)	(2,7)
3	(3,0)	(3,1)	(3,2)	(3,3)	(3,4)	(3,5)	(3,6)	(3,7)
4	(4,0)	(4,1)	(4,2)	(4,3)	(4,4)	(4,5)	(4,6)	(4,7)
5	(5,0)	(5,1)	(5,2)	(5,3)	(5,4)	(5,5)	(5,6)	(5,7)
6	(6,0)	(6,1)	(6,2)	(6,3)	(6,4)	(6,5)	(6,6)	(6,7)
7	(7,0)	(7,1)	(7,2)	(7,3)	(7,4)	(7,5)	(7,6)	(7,7)

Table 1.
Cyclotomic matrix of order 8.

cyclotomic numbers (i.e. equality of solutions of $(a, b)_{2l^2}$). In the next subsection, we will discuss which cyclotomic numbers are enough for the construction of the cyclotomic matrix. Thus it helps us to the faster computation of cyclotomic matrix.

3.2 Determination of equality relation of cyclotomic numbers

This subsection presents the procedure to determine the equality relation of cyclotomic numbers (i.e. the relation of pair of $(a, b)_{2l^2}$), which reduces the complexity of solutions of pair of $(a, b)_{2l^2}$ (see also [36]). For the determination of cyclotomic matrices, it is not necessary to obtain all $4l^4$ cyclotomic numbers of order $2l^2$. The minimum number of cyclotomic numbers required to determine all the cyclotomic numbers (i.e. required for construction of cyclotomic matrix) depends on the value of positive integer k on expressing prime $p = 2l^2k + 1$. By (2), if k is even, then

$$(a, b)_{2l^2} = (b, a)_{2l^2} = (a - b, -b)_{2l^2} = (b - a, -a)_{2l^2} = (-a, b - a)_{2l^2} = (-b, a - b)_{2l^2} \tag{5}$$

otherwise

$$\begin{aligned}(a, b)_{2l^2} &= \left(b + l^2, a + l^2\right)_{2l^2} = \left(l^2 + a - b, -b\right)_{2l^2} = \left(l^2 + b - a, l^2 - a\right)_{2l^2} \\ &= (-a, b - a)_{2l^2} = \left(l^2 - b, a - b\right)_{2l^2}.\end{aligned} \tag{6}$$

Thus by (5) and (6), cyclotomic numbers $(a, b)_{2l^2}$ of order $2l^2$ can be divided into various classes.

- $2|k$ and $l \neq 3$: In this case, (5) gives classes of singleton, three and six elements. $(0, 0)_{2l^2}$ form singleton class, $(-a, 0)_{2l^2}$, $(a, a)_{2l^2}$, $(0, -a)_{2l^2}$ form classes of three elements where $1 \leq a \leq 2l^2 - 1 \pmod{2l^2}$ and rest $4l^4 - 3 \times 2l^2 + 2$ of the cyclotomic numbers form classes of six elements.

- $2|k$ and $l = 3$: In this case, (5) divide cyclotomic numbers $(a, b)_{18}$ of order 18 into classes of singleton, second, three and six elements. $(0, 0)_{18}$ form singleton class, $(-a, 0)_{18}$, $(a, a)_{18}$, $(0, -a)_{18}$ form classes of three elements, where $1 \leq a \leq 17 \pmod{18}$, $(6, 12)_{18} = (12, 6)_{18}$ which is grouped into classes of two elements and rest $4l^4 - 3 \times 2l^2$ of the cyclotomic numbers form classes of six elements.

- $2\nmid k$ and $l \neq 3$: Using (6), once again we get classes of singleton, three and six elements. $\left(0, l^2\right)_{2l^2}$ forms singleton class, $(0, a)_{2l^2}$, $\left(a + l^2, l^2\right)_{2l^2}$, $\left(l^2 - a, -a\right)_{2l^2}$ form classes of three elements, where $0 \leq a \neq l^2 \leq 2l^2 - 1\left(\bmod 2l^2\right)$ and rest $4l^4 - 3 \times 2l^2 + 2$ of the cyclotomic numbers form classes of six elements.

- $2\nmid k$ and $l = 3$: In this situation, (6) partitions cyclotomic numbers $(a, b)_{18}$ of order 18 into classes of singleton, two, three and six elements. Here $(0, 9)_{18}$ form singleton class, $(0, a)_{18}$, $(a + 9, 9)_{18}$, $(9 - a, -a)_{18}$ form classes of three elements, where $0 \leq a \neq 9 \leq 17 \pmod{18}$, $(6, 3)_{18} = (12, 15)_{18}$ which is grouped into classes of two elements and rest $4l^4 - 3 \times 2l^2$ of the cyclotomic numbers form classes of six elements.

Algorithm 1 Equality relation of cyclotomic numbers.

```
START
Declare integer variable e, l, p, k, flag.
INPUT l, an odd prime and e = 2l^2
Declare an array of size e × e, where each element of array is 2 tuple structure
   (i.e. ordered pair of (a, b)_{2l^2}, where a and b are integers).
INPUT p, prime number greater than 2
if (p − 1)%e == 0 then
   k = (p − 1)/e
   if k even then
       Update table (E)
   else
       Update table (O)
   end if
end if
```

Here **Update table (E)** means each entry $(a, b)_{2l^2}$ of the table will be updated by applying the relations $(a, b)_{2l^2} = (b, a)_{2l^2} = (a - b, -b)_{2l^2} = (b - a, -a)_{2l^2} = (-a, b - a)_{2l^2} = (-b, a - b)_{2l^2}$, and **Update table (O)** means each entry $(a, b)_{2l^2}$ of the table will be updated by applying the relations $(a, b)_{2l^2} = (b + l^2, a + l^2)_{2l^2} = (l^2 + a - b, -b)_{2l^2} = (l^2 + b - a, l^2 - a_{2l^2}) = (-a, b - a)_{2l^2} = (l^2 - b, a - b)_{2l^2}$.

Further, if entries of the updated table are non-negative, then each entry should be replace by $(\text{mod}2l^2)$, otherwise add $2l^2$. It is clear from above exploration, cyclotomic numbers of order $2l^2$ are divided into different classes depending on the values of k and l. For $l = 2$ and let k be even, then $(0, 0)_8$ give unique solution, cyclotomic numbers of the form $(-a, 0)_8$, $(a, a)_8$, $(0, -a)_8$ where $1 \le a \le 7 \pmod 8$ gives the same solutions and rest of cyclotomic numbers (i.e. 42) which forms classes of six elements has maximum 7 distinct numbers of solutions. Therefore the initial table (i.e. **Table 1**) of cyclotomic matrix reduces to **Table 2**. Similarly, for $l = 2$ and let k be odd, then $(0, 4)_8$ give unique solution, cyclotomic numbers of the form $(0, a)_8$, $(a + 4, 4)_8$, $(4 - a, -a)_8$ where $0 \le a \ne 4 \le 7 (\text{mod}8)$ gives the same solutions and rest of cyclotomic numbers (i.e. 42) which forms classes of six elements has maximum 7 distinct numbers of solutions. Therefore the initial table

(a,b)				**b**				
a	**0**	**1**	**2**	**3**	**4**	**5**	**6**	**7**
0	(0,0)	(0,1)	(0,2)	(0,3)	(0,4)	(0,5)	(0,6)	(0,7)
1	(0,1)	(0,7)	(1,2)	(1,3)	(1,4)	(1,5)	(1,6)	(1,2)
2	(0,2)	(1,2)	(0,6)	(1,6)	(2,4)	(2,5)	(2,4)	(1,3)
3	(0,3)	(1,3)	(1,6)	(0,5)	(1,5)	(2,5)	(2,5)	(1,4)
4	(0,4)	(1,4)	(2,4)	(1,5)	(0,4)	(1,4)	(2,4)	(1,5)
5	(0,5)	(1,5)	(2,5)	(2,5)	(1,4)	(0,3)	(1,3)	(1,6)
6	(0,6)	(1,6)	(2,4)	(2,5)	(2,4)	(1,3)	(0,2)	(1,2)
7	(0,7)	(1,2)	(1,3)	(1,4)	(1,5)	(1,6)	(1,2)	(0,1)

Table 2.
Cyclotomic matrix of order 8 for even k.

(a,b)	b							
a	0	1	2	3	4	5	6	7
0	(0,0)	(0,1)	(0,2)	(0,3)	(0,4)	(0,5)	(0,6)	(0,7)
1	(1,0)	(1,1)	(1,2)	(1,3)	(0,5)	(0,3)	(1,3)	(1,7)
2	(2,0)	(2,1)	(2,0)	(1,7)	(0,6)	(1,3)	(0,2)	(1,2)
3	(1,1)	(2,1)	(2,1)	(1,0)	(0,7)	(1,7)	(1,2)	(0,1)
4	(0,0)	(1,0)	(2,0)	(1,1)	(0,0)	(1,0)	(2,0)	(1,1)
5	(1,0)	(0,7)	(1,7)	(1,2)	(0,1)	(1,1)	(2,1)	(2,1)
6	(2,0)	(1,7)	(0,6)	(1,3)	(0,2)	(1,2)	(2,0)	(2,1)
7	(1,1)	(1,2)	(1,3)	(0,5)	(0,3)	(1,3)	(1,7)	(1,0)

Table 3.
Cyclotomic matrix of order 8 for odd k.

(i.e. **Table 1**) of cyclotomic matrix reduces to **Table 3**. One can observe that 64 pairs of two parameter numbers $(a,b)_8$ reduced to 15 distinct pairs (see **Tables 2** and **3**).

Remark 3.0 By Algorithm 1, to compute $2l^2$ cyclotomic numbers, it is enough to compute $2l^2 + \lceil (2l^2-1)(2l^2-2)/6 \rceil$, if $(2l^2-1)(2l^2-2)|6$, otherwise $2l^2 + \lceil (2l^2-1)(2l^2-2)/6 \rceil + 1$. Further, when l is the least odd prime i.e. $l=3$, then $(2l^2-1)(2l^2-2) \nmid 6$. Therefore $l=3$, it is enough to calculate 64 distinct cyclotomic numbers of order $2l^2$ and for $l \neq 3$, it is sufficient to calculate $2l^2 + (2l^2-1)(2l^2-2)/6$ distinct cyclotomic numbers of order $2l^2$.

3.3 Determination of generators of $\mathbf{F}_p^*$

To determine the solutions of (1), we need the generator of the cyclic group $\mathbf{F}_p^*$. Let us choose finite field of order p that satisfy $p = 2l^2k + 1; k \in \mathbf{Z}^+$. Let $\gamma_1, \gamma_2, \gamma_3, \dots, \gamma_n$ be generators of $\mathbf{F}_p^*$. We consider finite field of order 17 (i.e. $\mathbf{F}_{17}$), since the chosen value of $p = 17$ with respect to the value of l take previously. Now to determine the generators of cyclic group $\mathbf{F}_{17}^*$. The detail procedure to obtain the generator of $\mathbf{F}_{17}^*$ has been depicted in Algorithm 2. If G_{17} is a set that contain all the generator of $\mathbf{F}_{17}^*$, we could get elements of G_{17} as $\{3, 5, 6, 7, 10, 11, 12, 14\}$.

Algorithm 2 Determination of generators of $\mathbf{F}_p^*$.

Declare integer variable p, count
Declare integer array $arr\mathbf{F}_p[p]$, $arr\mathbf{F}_p flag[p]$
for $i = 1$ to $p-1$ **do**
    $arr\mathbf{F}_p[i] = i$, $arr\mathbf{F}_p flag[i] = 0$
end for
Declare integer array $arr\mathbf{G}_p[max]$
Declare integer variable $flag = 0$, r, γ
for $i = 1$ to $p-1$ **do**
    count = 0
    for $f = 1$ to $p-1$ **do**
        $arr\mathbf{F}_p flag[f] = 0$
    end for
    $\gamma = arr\mathbf{F}_p[i]$

```
    for a = 1 to p − 1 do
      r = power(γ, a)   (mod p)
      for j = 1 to p − 1 do
        if r is equal to arrF_p[j] then
          arrF_pflag[j] = 1
        end if
      end for
    end for
    for k = 1 to p − 1 do
      if arrF_pflag[k] is equal to 1 then
        count++
      end if
    end for
    if count is equal to p − 1 then
      γ is generator
    end if
end for
```

3.4 Generation of cyclotomic matrices

This subsection, present an algorithm for the generation of cyclotomic matrices of order $2l^2$. Note that entries of cyclotomic matrices are solutions of (1). Thus we need the generator of the cyclic group $\mathbf{F}_p^*$, which is discussed in the previous subsection. On substituting the generators of $\mathbf{F}_p^*$ in Algorithm 3, we obtain the cyclotomic matrices of order $2l^2$ corresponding to different generators of $\mathbf{F}_p^*$. The chosen value of $p = 17$ implies $k = 2$ w.r.t. assume value of $l = 2$. Therefore the cyclotomic matrix will be obtain from **Table 2**. Let us choose a generator (say $\gamma_1 = 3$) from set G_{17}. On substituting $\gamma_1 = 3$ in Algorithm 3, it will generate cyclotomic matrix of order 8 over $\mathbf{F}_{17}$ w.r.t. chosen generator $\gamma_1 = 3$. Matrix B_0 show the corresponding cyclotomic matrix of order 8 w.r.t. chosen generator $3 \in \mathbf{F}_{17}^*$.

$$\mathbf{B}_0 = \begin{bmatrix} 0 & 0 & 0 & 0 & 0 & 0 & 1 & 0 \\ 0 & 0 & 0 & 0 & 1 & 0 & 1 & 0 \\ 0 & 0 & 1 & 1 & 0 & 0 & 0 & 0 \\ 0 & 0 & 1 & 0 & 0 & 0 & 0 & 1 \\ 0 & 1 & 0 & 0 & 0 & 1 & 0 & 0 \\ 0 & 0 & 0 & 0 & 1 & 0 & 0 & 1 \\ 1 & 1 & 0 & 0 & 0 & 0 & 0 & 0 \\ 0 & 0 & 0 & 1 & 0 & 1 & 0 & 0 \end{bmatrix}.$$

Algorithm 3 Generation of cyclotomic matrix.

```
1: INPUT: The value of p, l, γ
2: Declare an array arr[ROW][COL] (where elements are two tuple structure)
3: Declare integer variable p, l, k, γ, x, y, A, s, t, a, b, count = 0, p_1, p_2
4: for a equal to 0 to number of rows do
5:   for b equal to 0 to number of columns do
6:     for x is equal to 0 to k do
```

for y is equal to 0 to k **do**
$p_1 = 2l^2 * s + arr[a][b].l$
$p_2 = 2l^2 * t + arr[a][b].m$
$A = power(\gamma, p_1) + power(\gamma, p_2) + 1$
if $A \pmod{p}$ is equal to 0 **then**
$count++$
end if
end for
end for
$arr[a][b].n = count$
$count = 0$
end for
end for

Remark 3.1 It is noted that if we change the generator of $\mathbf{F}_p^*$, then entries of cyclotomic matrices get interchanged among themselves but their nature remains the same.

Remark 3.2 It is obvious that (by (4)) cyclotomic matrices of order $2l^2$ is always singular if the value of $k = 1$.

4. The public-key cryptosystem

In this section, we present the approach for designing a public key cryptosystem using cyclotomic matrices discussed in Section 3. The scheme employ matrices of order $2l^2$, whose entries are cyclotomic numbers of order $2l^2$. The public key is a non-trivial generator, say γ' of a set of generator in $\mathbf{F}_p^*$ along with p and l. The set of generator is obtain by Algorithm 2. The chosen public keys generate a cyclotomic matrix as of required order (i.e. order of $2l^2$) make use of Algorithm 3. Here, we define a trapdoor one-way function $\phi : \mathbf{F}_p^* \rightarrow \mathbf{F}_p^*$ as $\phi(r_0) = log_{\gamma'}(\gamma'')$; $r_0 \in \vec{N}$, γ', γ'' are non-trivial generators of $\mathbf{F}_p^*$. Thus, the secret key are the values of p, l, γ'' & r_0. To encrypt a message, define composition of matrix as $M_{2l^2}(A * B) \rightarrow M_{2l^2}(C)$, where A is a message block matrix, B is a cyclotomic matrix w.r.t. $\gamma' \in \mathbf{F}_p^*$ and C is the ciphertext matrix. Other way one can define $M_{2l^2}(B * A) \rightarrow M_{2l^2}(C)$. Therefore, the length of the ciphertext in CAC is equal to $2l^2$.

To decrypt a message, an algorithm is required to expand the secret keys provided by the secret values. Therefore, the Algorithm 4 is utilized for this purpose.

Algorithm 4 Secrete key expansion.

SECRET INPUT: The values of p, l, r_0 and γ''
Algorithm 1
Algorithm 2

The main purpose, to utilize the above algorithm is to construct a non-singular cyclotomic matrix of order $2l^2$ w.r.t. non-trivial generator γ'' ($\gamma'' \neq \gamma'$) in $\mathbf{F}_p^*$. Now to decrypt the message, we define inverse composition relation of matrices, which is $M_{2l^2}(C * Z) \rightarrow M_{2l^2}(A)$, where matrix Z is obtain by some efficient algebraic computation of matrix. Other way one can define $M_{2l^2}(Z * C) \rightarrow M_{2l^2}(A)$ respectively.

4.1 Determination of matrix Z

The following steps have been taken for the determination of matrix Z.

1. Determine the equality of cyclotomic matrix of order $2l^2$ corresponding to the secret values of p & l, which is perform by Algorithm 1.

2. Each entry of equality of cyclotomic matrix is multiplied by r_0.

3. Compute the inverse of equality of cyclotomic matrix generated in step 2.

4. Finally, on substitution the values of the generated cyclotomic matrix corresponding to γ'' to an inverse matrix in step 3.

The following two algorithms (i.e. Algorithm 5 & 6) are utilized to encrypt and decrypt a message in the proposed CAC, respectively.

Algorithm 5 Encryption.

1: Transfer the plain text (message) into its numerical value and store in matrix of order $2l^2$
2: PUBLIC INPUT: The values of p, l and γ'
3: Execute Algorithm 3
4: Check: Generated cyclotomic matrix in step 3 is non-singular
5: Cipher matrix: Multiply cyclotomic matrix and the matrix generated in step 1
6: Ciphertext: The corresponding text values of matrix generated in step 5

Algorithm 6 Decryption.

1: Input: The cipher matrix/ciphertext
2: Execute Algorithm 4
3: Each entries of equality of cyclotomic matrix (i.e. output matrix of Algorithm 1) is multiply by r_0. The entries of the generated matrix are pair of cyclotomic number
4: Compute the inverse of generated matrix in step 3 and substitute the value of each pair of cyclotomic number from generated matrix in step 2
5: Now multiply the cipher text matrix to generated matrix in step 4, we get back to the original plain text message.

4.2 Computational complexity of the CAC

In this section, we would validate the computational complexity of the proposed CAC. The computational complexity measures the amount of computational effort required, by the best as of now known techniques, to break a system [2]. However, it is exceptionally hard to demonstrate the computational complexity of public-key cryptosystems [2, 3]. For instance, if the public modulus of RSA is factored into its prime components, at that point the RSA is broken. Be that as it may, it is not demonstrated that breaking RSA is identical to factoring its modulus [41]. Here, we study the computational complexity of the CAC by providing arguments related to the inversion of the one-way function in CAC to a best known computational algorithm. The complexity of anonymous decryption could be understood as; if we assume that an attacker wants to recover the secret key by using all the information's available to them. Then they need to solve the discrete logarithm problem (DLP) to find the secret key followed by a number of steps described in Algorithm 6. Since, the one-way function is define analogous to discrete logarithm problem (DLP). However, although most mathematicians and computer scientists believe that the DLP is unsolvable [42]. The complexity of the DLP depends on the cyclic group. It is believed to be a hard problem for the multiplicative group of a finite field of large cardinality. Therefore even determining the very first step is nearly unsolvable.

If it is the case that somehow attacker manages to solve the DLP, then they have to determine Eq. (1) and calculate all the solutions corresponding to different pairs $(a, b)_{2l^2}$. Further, it is required to determine the relation matrix based on equality relation among the solutions of Eq. (1). Where entries of the relation matrix are the two-tuple structure of $(a, b)_{2l^2}$. Finally, entries of inverse of the relation matrix are required to replace through the implication of DLP.

Here we could observe the computational complexity as it increases with the value of p and $2l^2$. Therefore it is nearly impossible to determine the secret key for a large value of p and $2l^2$; hence uphold the secure formulation claim of the proposed work.

4.3 An example of the CAC

In this section, we provide an example for the proposed CAC. The example is designed according to guidelines described in Section 4. The main purpose of this example is to show the reliability of our cryptosystem. It is important to note that this example is non-viable for the proposed CAC, since the values of the parameters are too small.

Example 1 Let us consider $2l^2 = 8$ (i.e. $l = 2$) and $p = 17$. Suppose we want to send a message X whose numerical value store in matrix $\mathbf{A}$ of order 8.

$$\mathbf{A} = \begin{bmatrix} 2 & 3 & 5 & 9 & 8 & 0 & 2 & 1 \\ 1 & 5 & 9 & 2 & 9 & 3 & 0 & 5 \\ 2 & 1 & 3 & 2 & 5 & 6 & 8 & 7 \\ 5 & 3 & 0 & 7 & 8 & 7 & 3 & 1 \\ 4 & 2 & 3 & 1 & 9 & 8 & 7 & 3 \\ 0 & 9 & 2 & 3 & 5 & 6 & 8 & 9 \\ 1 & 0 & 2 & 9 & 6 & 7 & 9 & 8 \\ 9 & 1 & 3 & 2 & 4 & 4 & 5 & 6 \end{bmatrix}$$

We choose two distinct non-trivial generators of a set of generator in $\mathbf{F}_{17}^*$ (the set of generator is obtain by employing Algorithm 2), say $\gamma' = 11$ and $\gamma'' = 3$. Now, we evaluate the complex relation between these chosen generators, which can perform by DLP. One can write $3^7 = 11 \pmod{17}$. Consider that $r_0 = 7$. The public key is the public values $l = 2$, $p = 17$ & $\gamma' = 11$ and the private key is the secret values $l = 2$, $p = 17$, $r_0 = 7$ & $\gamma'' = 3$. The public values generated cyclotomic matrix of order 8 as required, which is

$$\mathbf{B}_3 = \begin{bmatrix} 0 & 0 & 1 & 0 & 0 & 0 & 0 & 0 \\ 0 & 0 & 0 & 1 & 0 & 1 & 0 & 0 \\ 1 & 0 & 0 & 0 & 0 & 0 & 0 & 1 \\ 0 & 1 & 0 & 0 & 1 & 0 & 0 & 0 \\ 0 & 0 & 0 & 1 & 0 & 0 & 0 & 1 \\ 0 & 1 & 0 & 0 & 0 & 0 & 1 & 0 \\ 0 & 0 & 0 & 0 & 0 & 1 & 1 & 0 \\ 0 & 0 & 1 & 0 & 1 & 0 & 0 & 0 \end{bmatrix}$$

Determinant of $\mathbf{B}_3$ is equal to 1, implies non-singular. Now we encrypt the message $\mathbf{A}$ by multiplying matrix $\mathbf{B}_3$ and $\mathbf{A}$, which is as follows:

$$\mathbf{C} = \mathbf{B}_3 \times \mathbf{A} = \begin{bmatrix} 2 & 1 & 3 & 2 & 5 & 6 & 8 & 7 \\ 5 & 12 & 2 & 10 & 13 & 13 & 11 & 10 \\ 11 & 4 & 8 & 11 & 12 & 4 & 7 & 7 \\ 5 & 7 & 12 & 3 & 18 & 11 & 7 & 8 \\ 14 & 4 & 3 & 9 & 12 & 11 & 8 & 7 \\ 2 & 5 & 11 & 11 & 15 & 10 & 9 & 13 \\ 1 & 9 & 4 & 12 & 11 & 13 & 17 & 17 \\ 6 & 3 & 6 & 3 & 14 & 14 & 15 & 10 \end{bmatrix}$$

The matrix $\mathbf{C}$ is a ciphertext matrix. To transmit the message, entries of the matrix converted into text. To decrypt the message, first, we expand the secret keys which are performed by Algorithm 4. It generates a non-singular cyclotomic matrix of order 8, which is shown by matrix $\mathbf{B}_0$. Now each entry of equality of cyclotomic matrix (i.e. output matrix of Algorithm 1) is multiplied by $r_0 = 7$. We get matrix D whose entries are pair of cyclotomic numbers.

$$\mathbf{D} = \begin{bmatrix} (0,0) & (0,7) & (0,6) & (0,5) & (0,4) & (0,3) & (0,2) & (0,1) \\ (0,7) & (0,1) & (1,2) & (1,6) & (1,5) & (1,4) & (1,3) & (1,2) \\ (0,6) & (1,2) & (0,2) & (1,3) & (2,4) & (2,5) & (2,4) & (1,6) \\ (0,5) & (1,6) & (1,3) & (0,3) & (1,4) & (2,5) & (2,5) & (1,5) \\ (0,4) & (1,5) & (2,4) & (1,4) & (0,4) & (1,5) & (2,4) & (1,4) \\ (0,3) & (1,4) & (2,5) & (2,5) & (1,5) & (0,5) & (1,6) & (1,3) \\ (0,2) & (1,3) & (2,4) & (2,5) & (2,4) & (1,6) & (0,6) & (1,2) \\ (0,1) & (1,2) & (1,6) & (1,5) & (1,4) & (1,3) & (1,2) & (0,7) \end{bmatrix}$$

Now compute the inverse of $\mathbf{D}$ and substitute the value from $\mathbf{B}_0$ to each pair of cyclotomic numbers. The matrix becomes

$$\mathbf{D}^* = \begin{bmatrix} -1 & 1 & 1 & -1 & -1 & 1 & -1 & 1 \\ 1 & 0 & 0 & 1 & 0 & 0 & 0 & -1 \\ 1 & 0 & 0 & 0 & 0 & 0 & 0 & 0 \\ -1 & 1 & 0 & -1 & 0 & 1 & -1 & 1 \\ -1 & 0 & 0 & 0 & 0 & 0 & 0 & 1 \\ 1 & 0 & 0 & 1 & 0 & -1 & 1 & -1 \\ -1 & 0 & 0 & -1 & 0 & 1 & 0 & 1 \\ 1 & -1 & 0 & 1 & 1 & -1 & 1 & -1 \end{bmatrix}$$

Finally, we obtain $\mathbf{D}^* \times \mathbf{C} = \mathbf{A}$.

5. The complexity of CAC

Time and space are usually prominent factors to establish the effectiveness of security solutions. In the before seen sections, we have established the computational difficulty to break the proposed work. Further, we would demonstrate the complexity of the solution in terms of worst-case running time.

The time complexity of Algorithm 1 in worst case is $\mathcal{O}(e^2)$. Since formation of matrix of order e and Update_Table() individually will take $\mathcal{O}(e^2)$. In algorithm 2, **for** loop in line number 9, 15, and 17 contributes $\mathcal{O}(e^3)$ in worst case. Since,

$$e = \frac{p-1}{k}$$

$$\Rightarrow e^3 = \left(\frac{p-1}{k}\right)^3 \equiv \left(\frac{p^3}{k^3}\right)$$

Since k is a positive integer, therefore when k attains its minimum value i.e. 1,

$$\frac{p^3}{k^3} \equiv p^3 \equiv e^3.$$

For any higher value of k, there is guarantee that

$$\frac{p^3}{k^3} < e^3.$$

Hence, we conclude that Algorithm 2 can take $\mathcal{O}(e^3)$ in worst case.

Similarly, in Algorithm 3, **for** loop in line number 4, 5, 6, 7 contributes $e.\ e.\ k.\ k$ or say $\mathcal{O}(e^2k^2)$ running time in worst case. Using similar analogy as in case of Algorithm 2, worst case complexity will be $\mathcal{O}(e^2)$.

5.1 Encryption

Encryption as expressed in Algorithm 5 constitutes of three logical divisions and the complexity of encryption would be the sum of the complexity of its part. The state divisions within are as follows;

1. Generating cyclotomic matrix

2. Checking the singularity of the cyclotomic matrix.

3. Multiplication of generated cyclotomic matrix and matrix corresponds to plain text.

Starting from the generation of the cyclotomic matrix, comprises the total complexity $\mathcal{O}(e^2)$ as stated earlier. Further, checking singularity involves the computation of determinants of the matrix of order e. In worst case computing determinant of a matrix of order n by fast algorithm [43] takes $\mathcal{O}(n^{2.373})$. Hence, singularity of the cyclotomic matrix of order e could be computed in $\mathcal{O}(e^{2.373})$ time. Finally, multiplication of cyclotomic matrix of order e and matrix corresponds to plain text of order e will take $\mathcal{O}(e^{2.3728639})$ time. Therefore, Complexity of Encryption would become $\mathcal{O}(e^2) + \mathcal{O}(e^{2.373}) + \mathcal{O}(e^{2.3728639}) \equiv \mathcal{O}(e^{2.373})$. Thus a polynomial time complexity seems to be quite worthwhile.

5.2 Decryption

Decryption as expressed in Algorithm 6 that include Algorithm 4 which sums the complexity of Algorithm 1 and 3, therefore takes $\mathcal{O}(e^2) + \mathcal{O}(e^2) \equiv \mathcal{O}(e^2)$ time. Further, multiplication of cyclotomic matrix of order e by a constant value r_0, therefore yield $\mathcal{O}(e^2)$ complexity. Likewise, inverse of a matrix of order n can be computed by a fast algorithm [43] in $\mathcal{O}(n^{2.373})$, therefore, inverse of generated matrix of order e could be computed in $\mathcal{O}(e^{2.373})$ time. Finally multiplication of two matrix of order e could be computed in $\mathcal{O}(e^{2.3728639})$ by best known algorithm [44] till date. Therefore, Complexity of decryption would be $\mathcal{O}(e^2) + \mathcal{O}(e^2) + \mathcal{O}(e^{2.373}) + \mathcal{O}(e^{2.3728639})$, which becomes $\mathcal{O}(e^{2.373})$.

6. Conclusion

In this chapter, we have introduced a secured asymmetric key cryptography model applying the principle of cyclotomic numbers over a finite field. Procedure to generate cyclotomic matrix along with public & private key have been presented, where the relation between the public & private key has acquired by discrete logarithm problem (DLP). Finally, a convincing argument to strengthen the claim has been presented followed by the method of encryption, decryption & a numerical example.

Acknowledgements

The authors are thankful and acknowledge the Central University of Jharkhand, Ranchi, Jharkhand for providing the necessary facilities to carry out this research.

Mathematics Subject Classification (2010)

11T22; 11T71; 94A60; 11T24

Author details

Md. Helal Ahmed[1], Jagmohan Tanti[1*] and Sumant Pushp[2]

1 Department of Mathematics, Central University of Jharkhand, Ranchi, India

2 Department of Computer Science and Technology, Tezpur University, Assam, India

*Address all correspondence to: jagmohan.t@gmail.com

References

[1] Diffie, W., Hellman, M.: New directions in cryptography. IEEE transactions on Information Theory **22** (6), 644–654 (1976)

[2] Miller, V.S.: Use of elliptic curves in cryptography. In: Conference on the theory and application of cryptographic techniques, pp. 417–426. Springer (1985)

[3] Wiener, M.J., Zuccherato, R.J.: Faster attacks on elliptic curve cryptosystems. In: International workshop on selected areas in cryptography, pp. 190–200. Springer (1998)

[4] Ahmad, J.I., Din, R., Ahmad, M. Analysis review on public key cryptography algorithms. International Journal of Electrical and Computer Engineering (IJECE) **12**(2), 447–454 (2018)

[5] Korzhik, V.I., Turkin, A.I.: Cryptanalysis of mcelieces public-key cryptosystem. In: Workshop on the Theory and Application of of Cryptographic Techniques, pp. 68–70. Springer (1991)

[6] Razborov, A.A., Rudich, S.: Natural proofs. Journal of Computer and System Sciences **55**(1), 24–35 (1997)

[7] Shirolkar, D., Katre, S.: Jacobi sums and cyclotomic numbers of order l2. Acta Arithmetica **1**(147), 33–49 (2011)

[8] Menezes, A.J., Katz, J., Van Oorschot, P.C., Vanstone, S.A.: Handbook of applied cryptography. CRC press (1996)

[9] ElGamal, T.: A public key cryptosystem and a signature scheme based on discrete logarithms. IEEE transactions on information theory **31** (4), 469–472 (1985)

[10] Bruce, S. Applied cryptography. 2nd John Wiley and Sons, Inc (1996)

[11] Koblitz, N., Menezes, A.J.: A survey of public-key cryptosystems. SIAM review **46**(4), 599–634 (2004)

[12] Ourivski, A.V., Johansson, T.: New technique for decoding codes in the rank metric and its cryptography applications. Problems of Information Transmission **38**(3), 237–246 (2002)

[13] Johnson, D., Menezes, A.: The elliptic curve digital signature algorithm (ecdsa) (1999) 21. Johnson, D., Menezes, A., Vanstone, S.: The elliptic curve digital signature algorithm (ecdsa). International journal of information security **1**(1), 36–63 (2001)

[14] Katre, S., Rajwade, A.: Complete solution of the cyclotomic problem in $\mathbf{F}q$ for any prime modulus $l, q = p^a$, $\mathrm{p} \equiv 1(\mathrm{mod}\,1)$. Acta Arithmetica **45**(3), 183–199 (1985)

[15] Zuccherato, R.: Using a pki based upon elliptic curve cryptography: Examining the benefits and difficulties. Entrust-Securing Digital Identities and Information (2003).

[16] De Win, E., Mister, S., Preneel, B., Wiener, M.: On the performance of signature schemes based on elliptic curves. In: International Algorithmic Number Theory Symposium, pp. 252–266. Springer (1998)

[17] Meier, A.V.: The elgamal cryptosystem (2005)

[18] Davida, G.I., Walter, G.G.: A public key analog gyptosystem. In: Advances in CryptologyEUROCRYPT87

[19] Loidreau, P.: Designing a rank metric based mceliece cryptosystem. In: International Workshop on Post-

Quantum Cryptography, pp. 142–152. Springer (2010)

[20] Pollard, J.M.: Factoring with cubic integers. In: The development of the number field sieve, pp. 4–10. Springer (1993)

[21] Lau, T.S.C., Tan, C.H.: A new technique in rank metric code-based encryption. Cryptography **2**(4), 32 (2018)

[22] Sidelnikov, V.M., Shestakov, S.O.: On insecurity of cryptosystems based on generalized reed-solomon codes. Discrete Mathematics and Applications **2**(4), 439–444 (1992)

[23] Stinson, D.R., Paterson, M.: Cryptography: theory and practice. CRC press (2018)

[24] Gabidulin, E.M.: Theory of codes with maximum rank distance. Problemy Peredachi Informatsii **21**(1), 3–16 (1985)

[25] Canteaut, A., Chabaud, F.: A new algorithm for finding minimum-weight words in a linear code: application to mceliece's cryptosystem and to narrow-sense bch codes of length 511. IEEE Transactions on Information Theory **44** (1), 367–378 (1998)

[26] Chabaud, F., Stern, J.: The cryptographic security of the syndrome decoding problem for rank distance codes. In: International Conference on the Theory and Application of Cryptology and Information Security, pp. 368–381. Springer (1996)

[27] Overbeck, R.: Structural attacks for public key cryptosystems based on gabidulin codes. Journal of Cryptology **21**(2), 280–301 (2008)

[28] Gibson, J.K.: Severely denting the gabidulin version of the mceliece public key cryptosystem. Designs, Codes and Cryptography **6**(1), 37–45 (1995)

[29] Gibson, K.: The security of the gabidulin public key cryptosystem. In: International Conference on the Theory and Applications of Cryptographic Techniques, pp. 212–223. Springer (1996)

[30] Park, C.S.: Improving code rate of mceliece's public-key cryptosystem. Electronics Letters **25**(21), 1466–1467 (1989)

[31] Gabidulin, E., Ourivski, A.: Modified gpt pkc with right scrambler. In: International workshop on coding and cryptography (Paris, 8-12 January 2001), pp. 233–242 (2001)

[32] Gabidulin, E.M., Ourivski, A.V., Honary, B., Ammar, B.: Reducible rank codes and their applications to cryptography. IEEE Transactions on Information Theory **49**(12), 3289–3293 (2003)

[33] McEliece, R.J.: A public-key cryptosystem based on algebraic coding theory. Coding Thv **4244**, 114–116 (1978)

[34] Le Gall, F.: Powers of tensors and fast matrix multiplication. In: Proceedings of the 39th international symposium on symbolic and algebraic computation, pp. 296–303 (2014)

[35] Acharya, V.V., Katre, S. Cyclotomic numbers of order 2l, l an odd prime. Acta Arithmetica **69**(1), 51–74 (1995)

[36] Ahmed, M.H., Tanti, J. Computation of jacobi sums and cyclotomic numbers with reduced complexity. Bulletin of Pure & Applied Sciences-Mathematics and Statistics **38** (1), 466–470 (2019)

[37] Ahmed, M.H., Tanti, J. Cyclotomic numbers and jacobi sums: A survey. In: Class Groups of Number Fields and Related Topics, pp. 119–140. Springer (2020)

[38] Ahmed, M.H., Tanti, J., Hoque, A. Complete solution to cyclotomy of order $2l^2$ with prime l. The Ramanujan Journal **53**(3), 529–550 (2020)

[39] Koblitz, N.: Elliptic curve cryptosystems. Mathematics of computation **48**(177), 203 209 (1987)

[40] Sidelnikov, V.M., Shestakov, S.O.: On encryption based on generalized reedsolomon codes. Diskretnaya Math **4**, 57–63 (1992)

[41] Goldwasser, S., Bellare, M.: Lecture notes on cryptography. Available: http://www.cs.ucsd.edu/users/mihir/papers/gb.html (2008)

[42] Schneier, B.: Applied cryptography: protocols, algorithms, and source code in C. john wiley & sons (2007)

[43] Aho, A.V., Hopcroft, J.E. The design and analysis of computer algorithms. Pearson Education India (1974).

[44] Lin, M.C., Chang, T.C., Fu, H.L.: Information rate of mceliece's public-key cryptosystem. Electronics Letters **26** (1), 16–18 (1990)

Chapter 6

Conversational Code Analysis: The Future of Secure Coding

Fitzroy Nembhard and Marco M. Carvalho

Abstract

The area of software development and secure coding can benefit significantly from advancements in virtual assistants. Research has shown that many coders neglect security in favor of meeting deadlines. This shortcoming leaves systems vulnerable to attackers. While a plethora of tools are available for programmers to scan their code for vulnerabilities, finding the right tool can be challenging. It is therefore imperative to adopt measures to get programmers to utilize code analysis tools that will help them produce more secure code. This chapter looks at the limitations of existing approaches to secure coding and proposes a methodology that allows programmers to scan and fix vulnerabilities in program code by communicating with virtual assistants on their smart devices. With the ubiquitous move towards virtual assistants, it is important to design systems that are more reliant on voice than on standard point-and-click and keyboard-driven approaches. Consequently, we propose MyCodeAnalyzer, a Google Assistant app and code analysis framework, which was designed to interactively scan program code for vulnerabilities and flaws using voice commands during development. We describe the proposed methodology, implement a prototype, test it on a vulnerable project and present our results.

Keywords: secure coding, virtual assistant, code analysis, static analysis

1. Introduction

Computing systems face serious threats from attackers on a day-to-day basis. Devices within a network could be targeted or used as launching pads to spawn malware and other attacks to critical systems and infrastructure. A system is as secure as its weakest link [1]. Therefore, software engineers must be cognizant of the cyber-related challenges that plague modern computer systems and engineer software with credible defenses. One of the first defenses against potential threats to computer systems is careful analysis of program code during development and taking necessary steps to minimize/eliminate vulnerabilities.

Program analysis falls into three main categories: static application security testing (SAST) or static analysis, dynamic application security testing (DAST) or dynamic analysis, and interactive application security testing (IAST). Static analysis is a "technique in which code listings, test results, or other documentation are ... examined ... to identify errors, violations of development standards, or other problems" [2]. Dynamic analysis is the "process of evaluating a system or component based on its behavior during execution" [2]. IAST involves instrumenting a program with sensors to monitor program code in memory during execution in order to find specific events that could cause vulnerabilities [3]. Two or more of these approaches may be combined to create hybrid tools and techniques for analyzing program code. These hybrid systems are designed to achieve more comprehensive coverage and to decrease the false positives and false negatives of existing approaches.

While researchers are interested in designing sound and complete code analysis tools, achieving soundness and completeness remains an intractable problem [4–6]. Consequently, a lot of research in code analysis is centered on improving the alerts of static analysis tools [4, 7]. More recently, several researchers have proposed models based on deep learning and other machine learning approaches to scan and fix vulnerabilities in program code [8]. Many of these tools are still at an infant stage and have not yet made it to market. Based on current trends, we believe that the future of code analysis will involve more refined tools based on artificial intelligence (AI), machine learning, and other hybrid approaches.

In this work, we propose a hybrid code analysis framework that employs the use of voice assistants (VAs) to allow a programmer to conversationally scan for and fix potential vulnerabilities in program code. The use of voice assistants have grown significantly in recent years. This work focuses primarily on the Google Assistant[1] as it is the most popular [9] among other virtual assistants.

The rest of the chapter is organized as follows: first, we discuss related work in the area of hybrid analysis in Section 2 followed by a discussion on challenges affecting adoption of existing approaches in Section 3. In Section 4, we theorize about the future of secure coding and propose a new code analysis approach in Section 5. We then use a case study to evaluate our proposed approach in Section 6 and present our conclusion in Section 7.

2. Related work

This work falls in the area of hybrid analysis. In this section, we summarize works in this area.

In 2006, Aggarwal and Jalote [10] combined static and dynamic analysis to detect buffer overflow in C programs. Both static and dynamic approaches have advantages and disadvantages. One of the disadvantages of dynamic analysis is the requirement of a large number of test cases, which present an overhead. Some dynamic analysis tools use a feature know as generate-and-patch or generate-and-validate in an effort to auto-fix vulnerabilities. In 2015, the authors of [11] analyzed reported patches for several DAST tools including GenProg, RSRepair, and AE, and found that the overwhelming majority of reported patches did not produce correct outputs. The authors attributed the poor performance of these tools to weak proxies (bad acceptance tests), poor search spaces that do not contain correct patches, and random genetic search that does not have a smooth gradient for the genetic search to traverse to find a solution [11].

In 2012, [12] proposed a hybrid approach that uses source code program slicing to reduce the size of C programs while performing analysis and test generation. The authors used a minimal slicing-induced cover and alarm dependencies to diminish the costly calls of dynamic analysis [13].

---

[1] Google, Google Assistant, and Dialogflow are registered trademarks of Google, Inc. The use of these names or tools and their respective logos are for research purposes and does not connote endorsement of this research by Google, Inc. or any of its partners.

In 2014, [14] implemented a hybrid architecture as the JSA analysis tool, which is integrated into the IBM AppScan Standard Edition product. The authors augmented static analysis with (semi-)concrete information by applying partial evaluation to JavaScript functions according to dynamic data recorded by the Web crawler. The dynamic component rewrites the program per the enclosing HTML environment, and the static component then explores all possible behaviors of the partially evaluated program.

In 2015, [15] applied a program slicing technique, similar to [12], to create a tool called *Flinder-SCA*. The authors also implemented their program using the *Frama-C* platform. The main difference between [12, 15] is that [15] performs abstract interpretation and taint analysis via a fuzzing technique wheres [12] does not perform taint analysis or fuzzing.

Also, in 2015, [16] proposed a hybrid malicious code detection scheme that was designed using an AutoEncoder and Deep Belief Networks (DBN). The AutoEncoder deep learning method was used to reduce the dimensionality of data. The DBN was composed of a multilayer Restricted Boltzmann Machines (RBM) and a layer of BP neural network. The model was tested on the KDDCUP'99 dataset but not on actual program code.

In 2019, [17] proposed SapFix, a static and dynamic analysis tool which combines a mutation-based technique, augmented by patterns inferred from previous human fixes, with a reversion-as-last resort strategy for fixing high-firing crashes. This tool is built upon Infer [18] and a localization infrastructure that aids developers in reviewing and fixing errors rapidly. Currently, SapFix is targeted at null pointer exception (NPE) crashes, but has achieved much success at Facebook [18].

In a dissertation produced in 2021, [19] proposed a code generation technique for Synchronous Control Asynchronous Dataflow (SCAD) processors based on a hybrid control-flow dataflow execution paradigm. The model is inspired by classical queue machines that completely eliminates the use of registers. The author uses satisfiability (SAT) solvers to aid in the code generation process [19].

To the best of our knowledge, our work is the first to employ modern virtual assistants to conversationally scan and fix vulnerabilities in program code. In [20], the authors established a voice user interface (VUI) for controlling laboratory devices and reading out specific device data. The results of their experiments produced benchmarks of established infrastructure and showed a high mean accuracy ($95\% \pm 3.62$) of speech command recognition and reveals high potential for future applications of a VUI within laboratories. In like manner, we propose the integration of personal assistants with code analysis systems to encourage programmers to produce more secure code.

3. Challenges affecting adoption of existing approaches

Several code analysis and vulnerability detection surveys have categorized tools in the literature [7, 21–23]. While surveys are essential in advancing research, many of them do not focus on tools found on websites. It must be noted that the average programmer does not look for tools in research papers. To that end, we conducted a Google search and found several popular websites that present various tools that programmers may use to scan their code for vulnerabilities. **Figure 1** shows a bar chart highlighting the number of tools found on these websites. As shown in the figure, GitHub and Wikipedia list the most tools and are often the top websites returned in search results due to their popularity. We further grouped the most popular static analysis tools found on these websites by language as shown in

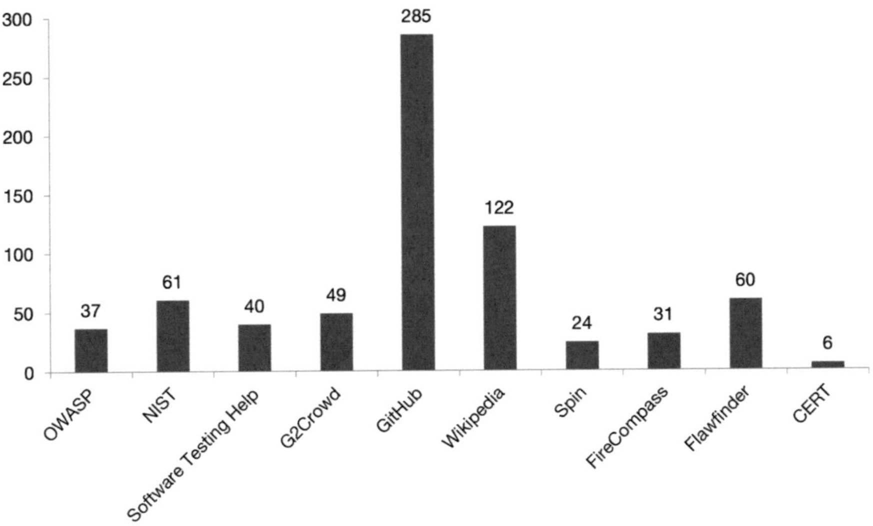

Figure 1.
The large number of code analysis tools found on popular websites.

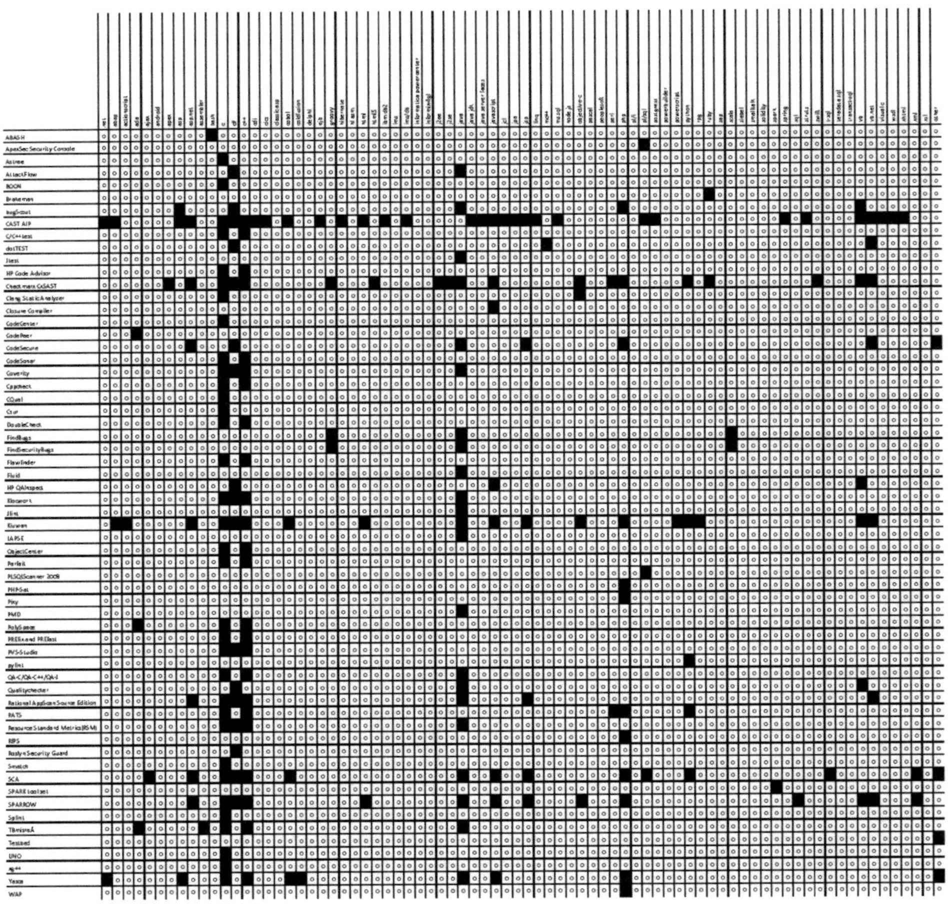

Figure 2.
Static analysis tools categorized by programming language.

Figure 2. As can be seen, this non-exhaustive list could overwhelm many programmers in determining the best tools for their projects.

In addition, the ability to combine code analysis approaches coupled with the number of programming languages that exist result in a large number of tools from which coders can choose to analyze their code. This makes it onerous for a programmer or organization to decide on a particular code analysis tool. Further, tools often require special configuration, which may take time to fine tune for best results. Many tools also suffer from usability issues, lengthy vulnerability reports, and false positives, making programmers avoid them altogether [24–26].

Another challenge affecting adoption of code analysis tools is monopolization of the market by certain companies. For-profit companies usually have the resources to improve tools by adding more state-of-the-art approaches such as cloud-based scanning, IAST support, and report generation. While these developments often advance the field of code analysis, they sometimes discourage small organizations and individuals from investing the effort and resources required to procure state-of-the-art tools. Thus, a streamlined, modern, cost-effective approach is needed to help encourage programmers to produce more secure code.

4. The future of code analysis

We believe that the future of code analysis lies in hybrid systems that combine several approaches to achieve useful analyses and actionable reports that will encourage programmers to produce more secure software. Based on current trends in machine learning, especially in deep learning, and natural language processing (NLP) (e.g., virtual assistants), it is safe to say that future code analysis will rely heavily on AI, ontologies, NLP, and machine learning. For example, when discussing the trends and challenges of machine learning, the authors in [27] "envision a fruitful marriage between classic logical approaches (ontologies) with statistical approaches which may lead to context-adaptive systems (stochastic ontologies) that might work similar to the human brain" [27].

Our projection is that code analysis frameworks will facilitate plug-and-play (PnP) models. **Figure 3** illustrates a generalized PnP model that uses virtual assistants to manage the analysis process. Using this plug-and-play model, programmers may select the code analyzer that best fits their project based on factors such as project type, project size, speed, efficiency, security, etc. This is similar to the

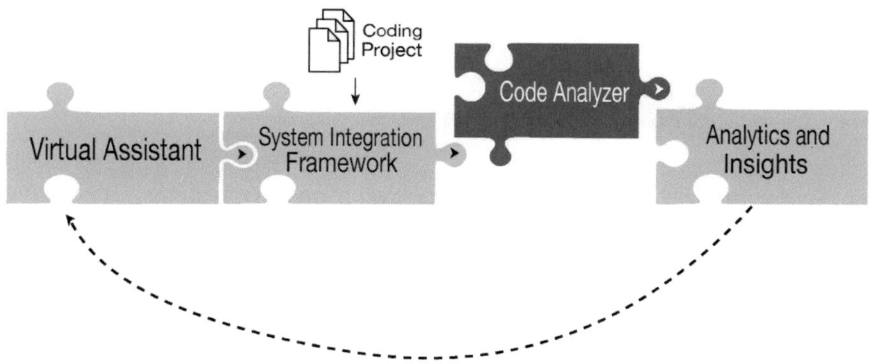

Figure 3.
A suggested model showing code analysis as part of a plug-and-play paradigm that facilitates the inclusion of any analysis tool and the use of a virtual assistant to manage the analysis process.

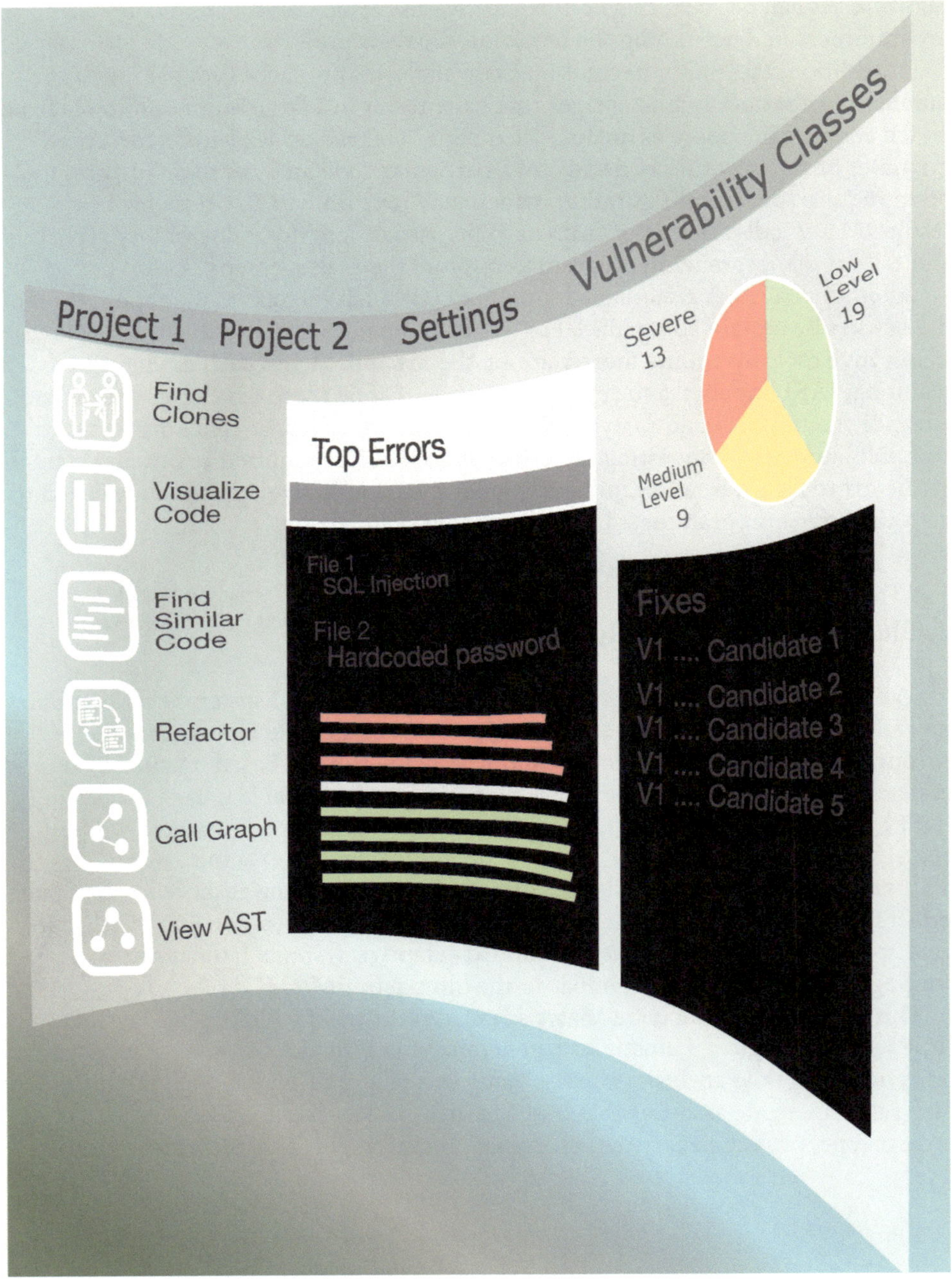

Figure 4.
A mockup of an analytical dashboard for code analysis on a curved display.

current landscape with virtual assistants and recommender systems. Currently, a person may use a virtual assistant like the Google Assistant to navigate a list of restaurants based on price, location, menu, reviews, etc. The virtual assistant may update the users preferences based on selections over time. This concept can also apply in code analysis where the chosen scanner used in the PnP model could be based on past scans or popularity.

The code analyzer featured in the model in **Figure 3** may use any combination of approaches including SAST, DAST, and IAST, which could be cloud-based or localized to the user's computer. These approaches could be backed by any algorithm that results in significant performance gains. It has been shown in the literature that deep learning and other ensemble methods perform very well in a large number of contexts including infected host detection [28], intrusion detection systems [29, 30], and malware analysis [31, 32], to name a few. Interestingly, many of these approaches can be used to create or improve code analyzers in an effort to help programmers produce more secure software.

Another feature of code analyzers of the future is a deep reliance on data analytics, visualizations and state-of-the-art interfaces. As discussed in the literature [8, 33], the interface of a code analyzer can have a negative or positive impact on its use and adoption. Therefore, for a system to be adopted in any project or organization, users must be able to gain insights from the way it presents its results. **Figure 4** shows a mockup of what we believe the interface of future code analyzers will look like. These interfaces will be in the form of dashboards instead of the customary lengthy bug reports displayed in a console.

5. Proposed approach

The proposed approach is to integrate a virtual assistant with a code analysis framework that allows users to scan, analyze, refactor and fix their code of inconsistencies and vulnerabilities. In this section, we describe the proposed approach using the system architecture.

5.1 System architecture

The system architecture for MyCodeAnalyzer is shown in **Figure 5**. The system consists of three main components: the virtual assistant, the webhook API and the

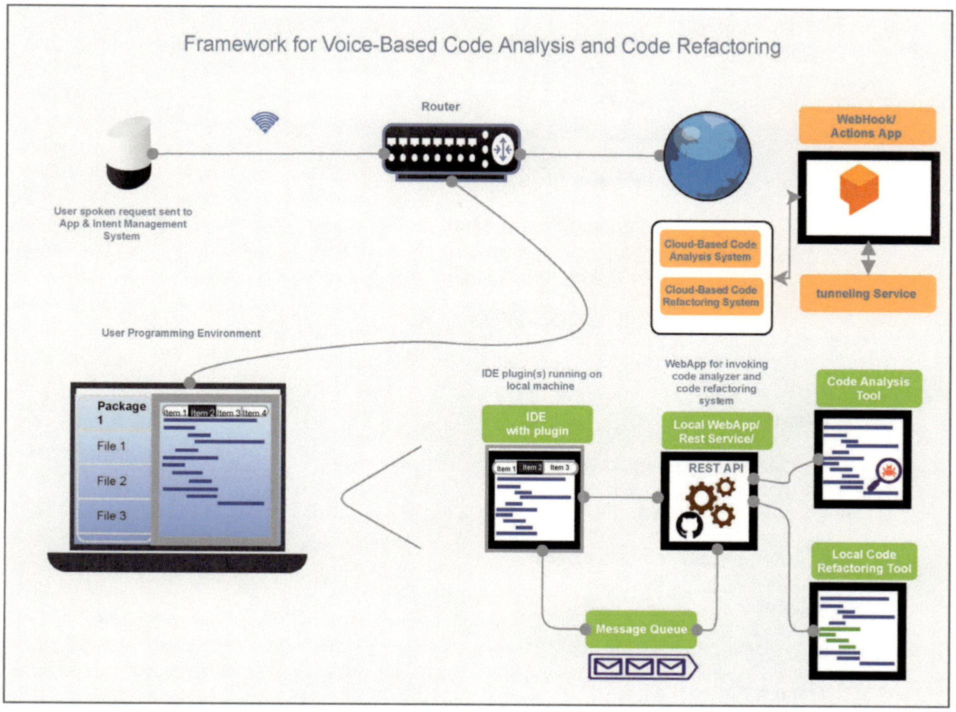

Figure 5.
MyCodeAnalyzer system architecture.

code scanning environment. The code scanning environment consists of a web app, an integrated development environment (IDE) plugin, code analyzers and refactoring tools. Google Assistant was chosen as the virtual assistant because of its popularity and easy-to-use App Engine and Dialogflow frameworks. The process flow is as follows: a user invokes a Google Assistant app (aka, Google Actions app) using a set of phrases understood by the system. This app is specially designed to understand trigger phrases associated with code analysis. Trigger phrases are training phrases that are entered into Dialogflow using an intent management system. Dialogflow is a natural language understanding platform that allows users to design and integrate a conversational user interface into a mobile app, web application, device, bot, interactive voice response system, etc. [34]. **Figure 6** captures the current intents incorporated into MyCodeAnalyzer. Each intent is backed by machine learning and NLP technology that uses named entity recognition (NER) and other approaches to extract entities from speech, determine context, and carry out tasks.

Figure 6.
Current Dialogflow intents used by MyCodeAnalyzer.

The intents in MyCodeAnalyzer are organized into 6 main categories: *Default Welcome Intent, vulnerability-scanning, clone-detection, Cancel, Bye, and Default Fallback Intent*. The *Default Welcome Intent* is used to welcome the user to the system and provide a description of potential requests that the application can fulfill. The *vulnerability-scanning* intent is the most complex of the intents and uses a tree-like structure to allow the user to conversationally scan a project for vulnerabilities, email a scan report or auto-fix errors based on the capabilities of the code analyzer. The *clone-detection* intent is used to scan a project for duplicated code and to provide a visualization showing a side-by-side comparison of similar code. While clones may not be vulnerable, they could become bloat in a project and could potentially lead to vulnerabilities. The *Cancel* intent is used to exit a task currently underway. *Bye* is used to exit the system and the *Default Fallback Intent*, as the name suggests, is used to ask the user to repeat a phrase for clarification or serve as a graceful fail mechanism.

Once invoked, the Google Assistant app communicates with the Google Conversation API to determine the user's intent. After intent has been determined, the Google Actions app then uses webhooks to communicate with a web service running on the user's computer. Using a tunneling service, the web service interacts with the user's IDE by way of a plugin. This plugin invokes a code analyzer or refactoring tool, takes actions based on the user's request, and places a message in a message queue. The web service then reads the queue and returns the message to the Google Assistant app, which then reads the message back to the user. The webhooks were set up in Dialogflow and run as servlets on Google App Engine. A servlet accepts valid Dialogflow POST requests and responds with data that is processed by the Google Assistant app and returned as output messages to the user. **Figure 7** further shows the internals of the system during a conversation between the user and the assistant. While only the static analysis portion of the system is demonstrated in this work, the system is modular enough for dynamic and hybrid analysis tools to be incorporated

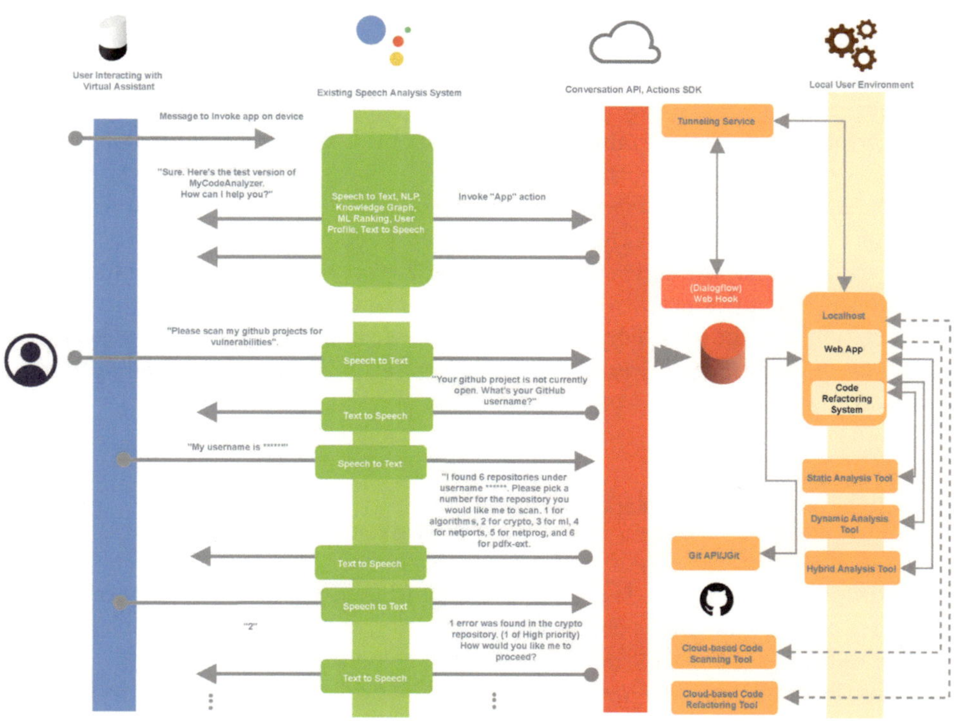

Figure 7.
Internals of MyCodeAnalyzer showing the flow of information throughout the system.

using the PnP approach discussed in Section 4. This approach provides a more complete code analysis depending on the user's preferences.

5.2 Accessing information about the coding environment

Two types of code-related information are accessed on the user's computer: code within the IDE and code from a Git repository (e.g., GitHub) currently opened in a web browser. The first type of information is important because it helps us to scan code being actively developed, while the second type is used in the case where the user would like to ensure that a repository is safe before forking it. MyCodeAnalyzer can detect GitHub pages that are open in a browser. On systems running MacOS, Applescript is used to communicate with the web browser. Other approaches will be employed in the future to reproduce this functionality on machines running other operating systems.

In order to access the user's computer to scan the code being worked on in the IDE or referenced in the browser, a methodology must be established to access this information in a minimally invasive manner. To do so, we created a plugin for a given IDE. Currently, we have plugins for IntelliJ IDEA and Eclipse. The plugin becomes a part of the IDE, monitors the code being developed, and updates a message queue (data file) with information about the code files and projects manipulated by the programmer. Also, special system calls are used to access any browser tabs that point to GitHub projects. A local web app in the form of a Spring MVC REST API [35] runs on the user' s computer. The job of the local web app is to communicate with MyCodeAnalyzer by way of a tunnel in order to scan local code or GitHub projects displayed in the user' s web browser.

5.2.1 Accessing code within the IDE

Listing 1 shows the Applescript code that is used to check for gui-based applications that are currently open on the user' s computer. Following this is a snapshot of the corresponding output, which includes the Intellij IDEA IDE in the list. This Applescript code is added to the REST app where it is run on localhost and invoked by MyCodeAnalyzer to determine if the user is actively using an IDE. To further contextualize the process of determining which code the user would like to scan, it is also of interest to find out the *frontmost* or most active application on the user's computer. To do so, the code shown in Listing 2 was used. This code is expected to return a single application, which in turn allows MyCodeAnalyzer to return a more direct response to the user. For example, a response might be, "*Say IDE, if you would like me to scan the code that you are currently working on in IntelliJ*" instead of using indirect phrases such as "*... may be working on.*"

```
set text item delimeters to ", "
tell application "System Events" to
(name of every process where background only is false) as text end tell
```

Listing 1. Applescript code used to list all gui-based applications that are currently running on the user' s computer.

The following is a sample output generated using the code in Listing 1:

```
"Google Chrome, Sublime Text, Terminal, idea, pycharm,Teams, Mail, teXShop,
Notes, Spotify, Finder, Microsoft PowerPoint, X11.bin, AdobeReader, iTunes,
Microsoft Excel,Script Editor, Activity Monitor, System Preferences, Safari,
Preview"
```

Since most IDEs are standalone applications, we believe the best way to have access to the user' s code in a minimally invasive manner is to be an "insider" (That is, to use a plugin that becomes part of the IDE). Consequently, the goal of the plugins was to monitor the code being developed by taking note of the coding project and the coding files being manipulated by the user. To accomplish this, listeners were added to the IDE to detect when the text editor portion of the IDE is active, when tabs are activated or switched, and when code files are edited. The message queue is updated with the following pieces of information when the aforementioned actions are performed: *ProjectName, ProjectLocation, CurrentFile, DateAdded, CurrentlyActive*. This queue is then queried for active files and projects when POST requests are made by the Google Assistant app to the local REST service running on the user's computer.

```
tell application "System Events"
name of application processes whose frontmost is true end tell
```

Listing 2. Applescript code used to determine the most active application on a computer.

5.2.2 Accessing code referenced by tabs opened in the web browser

Like IDEs, web browsers provide little to no way for outside tools to access their core areas. However, the Applescript-based techniques used previously for accessing the System Events utility can be used to access the tabs that are currently open in the web browser on the user' s device. Listing 3 is used to retrieve tabs currently open in Google Chrome. This script can be modified to get tabs in other browsers such as FireFox or Safari. MyCodeAnalyzer then checks if any of the URLs point to valid public GitHub accounts, which are then searched for coding projects if the user requests that a scan of a Git project be performed.

```
set text item delimeters to ","
tell application "Google Chrome" to URL of tabs of every window
as text
end tell
```

Listing 3. Applescript code used to retrieve tabs currently open in Google Chrome.

6. Case study

In this section, we present a case study that demonstrates an implementation of our proposed methodology. The main goal of this case study is to demonstrate the applicability of integrating a virtual assistant into a code analysis framework to allow the user to conversationally scan their code for vulnerabilities. The system is currently in a prototypical stage. Here we perform a scan of a coding project using the Google Assistant app via an Apple iPhone.

The following was done based on the proposed approach discussed in Section 5:

1. Create a Google Assistant app

   A Google Assistant app was created based on the intents depicted in **Figure 7**. Dialogflow, Google App Engine, and Google Actions Console are key components in the design of the app. Once designed, the app was tested using the Google Actions API Simulator as well as released in alpha mode and tested on a smart phone running the Google Assistant.

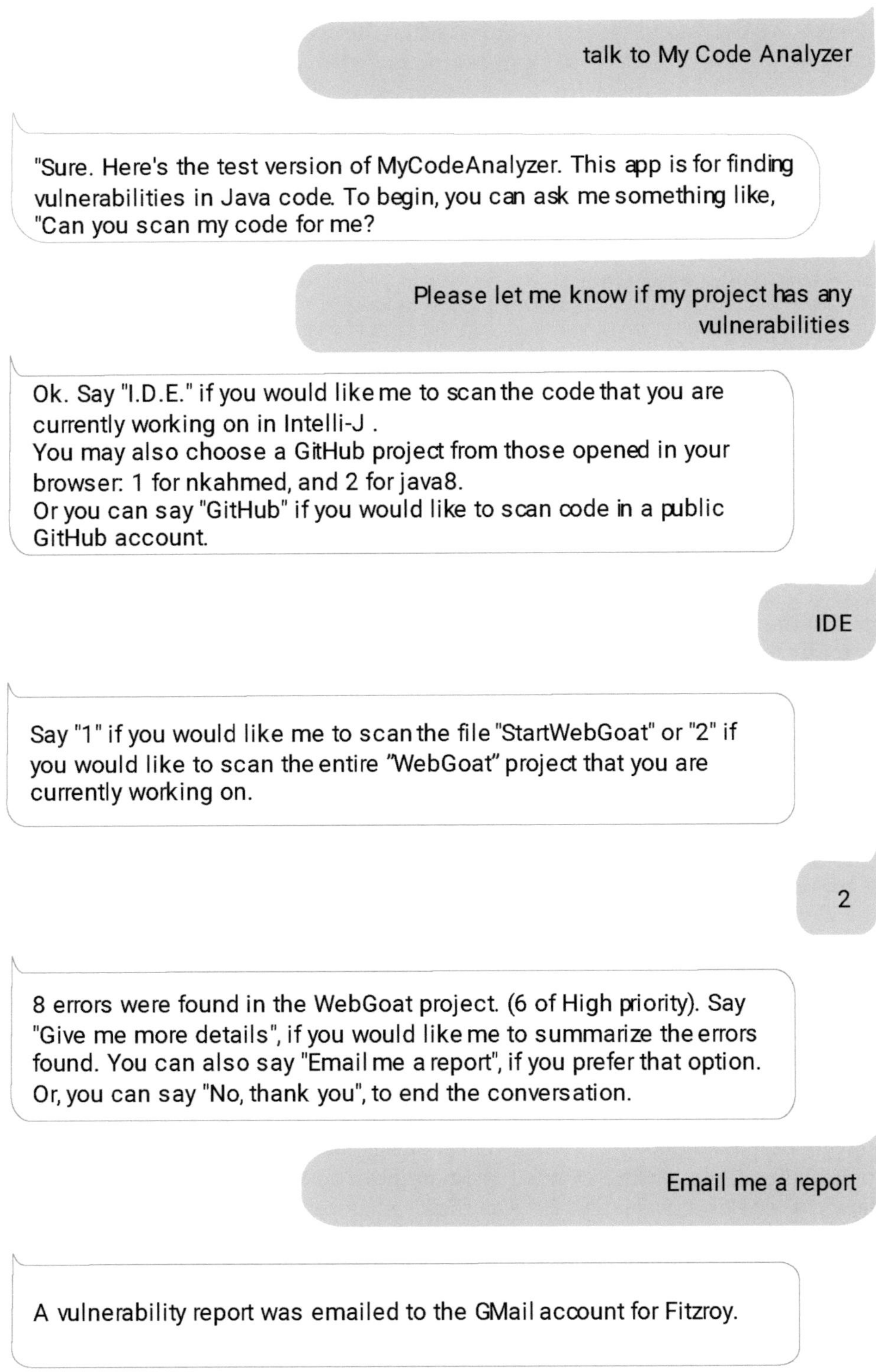

Figure 8.
A conversation between MyCodeAnalyzer and a human tester while scanning the OWASP WebGoat project.

2. Create a local web app to interface with the Google Assistant app and the coding environment

   The local web app was created using Spring Boot [35] and was launched on the computer via Apache Tomcat [36].

3. Create an IDE plugin for IntelliJ IDEA

   Our IntelliJ IDEA plugin was created and installed in IntelliJ version 2020.3.2. The plugin was installed using the IntelliJ plugin installer, which installs a local plugin from a JAR (Java ARchive) file.

4. Choose and integrate a code analyzer

   PMD [37] static code analyzer (version 6.31.0) was chosen for this study. PMD uses a rule-based system to find common programming flaws in code written in 8 programming languages, offering the most support for Java and Apex. The rules used by PMD are divided into categories such as best practices, error prone, and security. For this case study, a set of rules was selected from the error prone and security categories.

5. Chose a vulnerable project

   The OWASP WebGoat [38] project was used to evaluate the system. WebGoat is an insecure application that allows researchers and developers to test vulnerabilities commonly found in Java-based applications that use common and popular open source components [38].

6. Test the system and report results

   To integrate the Google Actions app with the local web app, Ngrok [39] was chosen as the tunneling tool. Ngrok is a tool that exposes local servers behind NATs and firewalls to the public Internet over secure tunnels [39].

6.1 Results and discussion

In this section, we capture a conversation between the Google Assistant app during the analysis of the WebGoat Project, present the report generated by the assistant, and discuss the results. It must be noted that the errors found by the Assistant during the code analysis are the same as those that would be produced by the standalone PMD project.

At this early stage of the project, the main benefit of the system is the ability to use a virtual assistant to perform code analysis while multitasking, thus improving productivity. After the system is setup, the programmer can configure and engage with the VA by voice without having to manually configure the code analyzer or browse and try to understand lengthy bug reports. The assistant can be used to perform actions based on the severity of the vulnerabilities found in the project. In the current version of MyCodeAnalyzer, Google Assistant can email the user a well-formatted report or read out the most important action items after analyzing the code. **Figure 8** captures a conversation between a human tester and the Google Assistant. **Figure 9** shows a formatted vulnerability report generated by the assistant and emailed to the user after scanning the WebGoat project. The WebGoat project has more severe vulnerabilities, but only those in the figure were captured by PMD based on the rulesets used by the analyzer. As can be seen from the report, MyCodeAnalyzer was able to process the lengthy XML reported returned by PMD

Vulnerability Scan Report

Below is the vulnerability scan report for your WebGoat Project

Problem	Class	Method	Priority	Line	Desc
Constructor calls overridable method	UserSessionData	UserSessionData	High	16	Overridable method 'setValue' called during object construction
Constructor calls overridable method	WebGoatUser	WebGoatUser	High	39	Overridable method 'createUser' called during object construction
Avoid branching statement as last in loop	JWTLessonTest	getSecretToken	Medium High	72	Avoid using a branching statement as the last in a loop.
Constructor calls overridable method	MD5	MD5	High	39	Overridable method 'reset' called during object construction
Return empty array rather than null	JWTFinalEndpoint	resolveSigningKeyBytes	High	164 - 175	Return an empty array rather than null.
Avoid branching statement as last in loop	JWTFinalEndpoint	resolveSigningKeyBytes	Medium High	169	Avoid using a branching statement as the last in a loop.
Constructor calls overridable method	DisplayUser	DisplayUser	High	54	Overridable method 'genUserHash' called during object construction
Constructor calls overridable method	WebGoatUser	WebGoatUser	High	61	Overridable method 'createUser' called during object construction

This report was generated by MyCodeAnalyzer. Vulnerability scan was conducted on 2021-02-20T20:57Feb 20, 2021 8:57:15 PM net.sourceforge.pmd.PMD encourageToUseIncrementalAnalysis WARNING: This analysis could be faster, please consider using Incremental Analysis: https://pmd.github.io/pmd-6.31.0/pmd_userdocs_incremental_analysis.html :15.921

Figure 9.
The report generated by MyCodeAnalyzer and emailed to the user after scanning the OWASP WebGoat project.

into a more easily understood report that captures only pertinent information. These results demonstrate the applicability of using a framework backed by virtual assistants to scan code for vulnerabilities and generate meaningful reports.

6.2 Challenges

It is important to outline some challenges with the use of VAs for code analysis and mitigation of vulnerabilities. The main challenge with this new approach to code analysis is adoption. A recent study involving a small sample of participants shows that currently the primary use of VAs are for music procurement (40% of users), for information (17%), and automation (9%) [40]. Since this is a new avenue of research, there may be initial challenges with adoption in the code analysis arena. However, we believe that as the market grows and coders get exposed to this technology, the adoption rates will increase. Researchers predict a growing use for digital voice assistants over the next few years [41, 42].

Another challenge with using the PnP model discussed in this research is handling the differences between output reports from different code analyzers. To mitigate this issue, the code analysis community may require standardization of vulnerability reports in popular formats such as XML, JSON, and HTML. Currently, most tools include information such as files, classes, and line numbers where errors are found. While the output formats may be different, NLP techniques such as NER can also be used to mine these reports for key pieces of information to achieve a standard format that can be handled by the virtual assistant and the proposed analysis framework.

7. Conclusion

Getting programmers to write secure code remains a challenge. Security is often sacrificed in an effort to add a feature to a software product or to meet a deadline. When security is sacrificed for other gains, the end result is a product riddled with bugs or vulnerabilities. Steps must be taken to encourage programmers to produce more secure software. In this research, we discussed the limitations of existing code analysis approaches and propose a framework that allows programmers to use virtual assistants to conversationally scan and fix potential vulnerabilities in their code. Virtual assistants are becoming popular in everyday activities such as procuring and listening to music, finding places of interest, managing a smart home, shopping, etc. We posit that as they become more mainstream, they can be used to manage code analysis while keeping programmers productive. We implement our proposed methodology using the Google Assistant and demonstrate its utility in an effort to find new, creative ways to help programmers produce more secure software. Future work will involve extending the model to use any applicable code analyzer based on a plug-and-play paradigm, adding data analytics and visualizations to help programmers draw insights from their code, implementing the refactoring and auto-fixing modules, and conducting a user study to evaluate the framework.

Abbreviations

DAST	Dynamic application security testing
IAST	Interactive application security testing
NLP	Natural language processing
PnP	Plug-and-play
SAST	Static application security testing
SCAD	Synchronous control asynchronous dataflow

Author details

Fitzroy Nembhard* and Marco M. Carvalho
Florida Institute of Technology, Melbourne, Florida, USA

*Address all correspondence to: fitzroy@ieee.org

References

[1] Bruce Schneier. Secrets & lies: Digital security in a networked world new york: Wiley computer publishing. 2000. *Ch*, 16:245–246.

[2] Iso/iec/ieee international standard - systems and software engineering–vocabulary. *ISO/IEC/IEEE 24765:2017 (E)*, pages 1–541, 2017. doi: 10.1109/IEEESTD.2017.8016712.

[3] F. Nembhard, M. Carvalho, and T. Eskridge. A hybrid approach to improving program security. In *2017 IEEE Symposium Series on Computational Intelligence (SSCI)*, pages 1–8, 2017. doi: 10.1109/SSCI.2017.8285247.

[4] Sarah Heckman and Laurie Williams. A systematic literature review of actionable alert identification techniques for automated static code analysis. *Information and Software Technology*, 53(4):363 – 387, 2011. Special section: Software Engineering track of the 24th Annual Symposium on Applied Computing.

[5] B. Chess and G. McGraw. Static analysis for security. *IEEE Security Privacy*, 2(6):76–79, Nov 2004.

[6] Brian Chess and Jacob West. *Secure programming with static analysis*. Pearson Education, 2007.

[7] T. Muske and A. Serebrenik. Survey of approaches for handling static analysis alarms. In *2016 IEEE 16th International Working Conference on Source Code Analysis and Manipulation (SCAM)*, pages 157–166, Oct 2016.

[8] Fitzroy D. Nembhard, Marco M. Carvalho, and Thomas C. Eskridge. Towards the application of recommender systems to secure coding. *EURASIP Journal on Information Security*, 2019(1):9, 2019. doi: 10.1186/s13635-019-0092-4. URL https://doi.org/10.1186/s13635-019-0092-4.

[9] Andreas M Klein, Andreas Hinderks, Maria Rauschenberger, and Jörg Thomaschewski. Exploring voice assistant risks and potential with technology-based users. In *Proceedings of 16th International Conference on Web Information Systems and technology (WEBIST)*, pages 1–8, 2020.

[10] A. Aggarwal and P. Jalote. Integrating static and dynamic analysis for detecting vulnerabilities. In *30th Annual International Computer Software and Applications Conference (COMPSAC'06)*, volume 1, pages 343–350, Sept 2006.

[11] Zichao Qi, Fan Long, Sara Achour, and Martin Rinard. An analysis of patch plausibility and correctness for generate-and-validate patch generation systems. In *Proceedings of the 2015 International Symposium on Software Testing and Analysis*, pages 24–36. ACM, 2015.

[12] Omar Chebaro, Nikolai Kosmatov, Alain Giorgetti, and Jacques Julliand. Program slicing enhances a verification technique combining static and dynamic analysis. In *Proceedings of the 27th Annual ACM Symposium on Applied Computing*, SAC '12, pages 1284–1291, New York, NY, USA, 2012. ACM.

[13] Fitzroy Nembhard. *A Recommender System for Improving Program Security Through Source Code Mining and Knowledge Extraction*. PhD thesis, Florida Institute of Technology, 2018.

[14] Omer Tripp, Pietro Ferrara, and Marco Pistoia. Hybrid security analysis of web javascript code via dynamic partial evaluation. In *Proceedings of the 2014 International Symposium on Software Testing and Analysis*, pages 49–59, 2014.

[15] Nir Piterman, editor. *Hardware and Software: Verification and Testing: 11th*

International Haifa Verification Conference, HVC 2015, Haifa, Israel, November 17-19, 2015, Proceedings. Springer International Publishing, Cham, 2015.

[16] Yuancheng Li, Rong Ma, and Runhai Jiao. A hybrid malicious code detection method based on deep learning. *International Journal of Security and Its Applications*, 9(5):205–216, 2015.

[17] A. Marginean, J. Bader, S. Chandra, M. Harman, Y. Jia, K. Mao, A. Mols, and A. Scott. Sapfix: Automated end-to-end repair at scale. In *2019 IEEE/ACM 41st International Conference on Software Engineering: Software Engineering in Practice (ICSE-SEIP)*, pages 269–278, 2019. doi: 10.1109/ICSE-SEIP.2019.00039.

[18] Cristiano Calcagno, Dino Distefano, Jeremy Dubreil, Dominik Gabi, Pieter Hooimeijer, Martino Luca, Peter O'Hearn, Irene Papakonstantinou, Jim Purbrick, and Dulma Rodriguez. Moving fast with software verification. In Klaus Havelund, Gerard Holzmann, and Rajeev Joshi, editors, *NASA Formal Methods*, pages 3–11, Cham, 2015. Springer International Publishing.

[19] Anoop Bhagyanath. *Code Generation for Synchronous Control Asynchronous Dataflow Architectures*. PhD thesis, Technical University of Kaiserslautern, 2021.

[20] Jonas Austerjost, Marc Porr, Noah Riedel, Dominik Geier, Thomas Becker, Thomas Scheper, Daniel Marquard, Patrick Lindner, and Sascha Beutel. Introducing a virtual assistant to the lab: A voice user interface for the intuitive control of laboratory instruments. *SLAS TECHNOLOGY: Translating Life Sciences Innovation*, 23(5):476–482, 2018.

[21] T. Muske and A. Serebrenik. Survey of approaches for handling static analysis alarms. In *2016 IEEE 16th International Working Conference on Source Code Analysis and Manipulation (SCAM)*, pages 157–166, 2016. doi: 10.1109/SCAM.2016.25.

[22] Anjana Gosain and Ganga Sharma. Static analysis: A survey of techniques and tools. In Durbadal Mandal, Rajib Kar, Swagatam Das, and Bijaya Ketan Panigrahi, editors, *Intelligent Computing and Applications*, pages 581–591, New Delhi, 2015. Springer India.

[23] G. Lin, S. Wen, Q. L. Han, J. Zhang, and Y. Xiang. Software vulnerability detection using deep neural networks: A survey. *Proceedings of the IEEE*, 108(10): 1825–1848, 2020. doi: 10.1109/JPROC.2020.2993293.

[24] F. Nembhard and M. Carvalho. The impact of interface design on the usability of code analyzers. In *2019 SoutheastCon*, pages 1–6, 2019. doi: 10.1109/SoutheastCon42311.2019.9020339.

[25] Brittany Johnson, Yoonki Song, Emerson Murphy-Hill, and Robert Bowdidge. Why don't software developers use static analysis tools to find bugs? In *Proceedings of the 2013 International Conference on Software Engineering*, ICSE '13, pages 672–681, Piscataway, NJ, USA, 2013. IEEE Press. ISBN 978-1-4673-3076-3.

[26] Ted Kremenek, Ken Ashcraft, Junfeng Yang, and Dawson Engler. Correlation exploitation in error ranking. In *ACM SIGSOFT Software Engineering Notes*, SIGSOFT '04/FSE-12, pages 83–93, New York, NY, USA, 2004. ACM. doi: 10.1145/1029894.1029909.

[27] Andreas Holzinger, Peter Kieseberg, Edgar Weippl, and A. Min Tjoa. Current advances, trends and challenges of machine learning and knowledge extraction: From machine learning to explainable ai. In Andreas Holzinger, Peter Kieseberg, A Min Tjoa, and Edgar Weippl, editors, *Machine Learning and*

Knowledge Extraction, pages 1–8, Cham, 2018. Springer International Publishing.

[28] Paula Venosa, Sebastian Garcia, and Francisco Javier Diaz. A better infected hosts detection combining ensemble learning and threat intelligence. In Patricia Pesado and Marcelo Arroyo, editors, *Computer Science – CACIC 2019*, pages 354–365, Cham, 2020. Springer International Publishing.

[29] Ngoc Tu Pham, Ernest Foo, Suriadi Suriadi, Helen Jeffrey, and Hassan Fareed M Lahza. Improving performance of intrusion detection system using ensemble methods and feature selection. In *Proceedings of the Australasian Computer Science Week Multiconference*, ACSW '18, New York, NY, USA, 2018. Association for Computing Machinery. ISBN 9781450354363. doi: 10.1145/3167918.3167951. URL https://doi.org/10.1145/3167918.3167951.

[30] S. A. Ludwig. Intrusion detection of multiple attack classes using a deep neural net ensemble. In *2017 IEEE Symposium Series on Computational Intelligence (SSCI)*, pages 1–7, 2017. doi: 10.1109/SSCI.2017.8280825.

[31] Bojan Kolosnjaji, Apostolis Zarras, George Webster, and Claudia Eckert. Deep learning for classification of malware system call sequences. In *Australasian Joint Conference on Artificial Intelligence*, pages 137–149. Springer, 2016.

[32] Daniel Gibert, Carles Mateu, and Jordi Planes. The rise of machine learning for detection and classification of malware: Research developments, trends and challenges. *Journal of Network and Computer Applications*, 153:102526, 2020. ISSN 1084-8045. doi: https://doi.org/10.1016/j.jnca.2019.102526. URL https://www.sciencedirect.com/science/article/pii/S1084804519303868.

[33] F. Nembhard and M. Carvalho. The impact of interface design on the usability of code analyzers. In *2019 SoutheastCon*, pages 1–6, 2019. doi: 10.1109/SoutheastCon42311.2019.9020339.

[34] Dialogflow, 2021. URL https://cloud.google.com/dialogflow/docs. Accessed: 2021-02-19.

[35] Spring boot, 2021. URL https://spring.io/projects/spring-boot. Accessed: 2021-02-19.

[36] Apache tomcat, 2021. URL http://tomcat.apache.org/. Accessed: 2021-02-19.

[37] PMD. Pmd source code analyzer project, 2021. URL https://pmd.github.io/. Accessed: 2021-02-18.

[38] Owasp webgoat, 2021. URL https://owasp.org/www-project-webgoat/. Accessed: 2021-02-19.

[39] ngrok. Ngrok, 2021. URL https://ngrok.com/. Accessed: 2021-02-19.

[40] Frank Bentley, Chris Luvogt, Max Silverman, Rushani Wirasinghe, Brooke White, and Danielle Lottridge. Understanding the long-term use of smart speaker assistants. *Proceedings of the ACM on Interactive, Mobile, Wearable and Ubiquitous Technologies*, 2 (3):1–24, 2018.

[41] Sven Tuzovic and Stefanie Paluch. Conversational commerce–a new era for service business development? In *Service business development*, pages 81–100. Springer, 2018.

[42] Andreas M Klein, Andreas Hinderks, Maria Rauschenberger, and Jörg Thomaschewski. Exploring voice assistant risks and potential with technology-based users. In *Proceedings of 16th International Conference on Web Information Systems and technology (WEBIST)*, pages 1–8, 2020.

Chapter 7

Non Classical Structures and Linear Codes

Surdive Atamewoue Tsafack

Abstract

This chapter present some new perspectives in the field of coding theory. In fact notions of fuzzy sets and hyperstructures which are consider here as non classical structures are use in the construction of linear codes as it is doing for fields and rings. We study the properties of these classes of codes using well known notions like the orthogonal of a code, generating matrix, parity check matrix and polynomials. In some cases particularly for linear codes construct on a Krasner hyperfield we compare them with those construct on finite field called here classical structures, and we obtain that linear codes construct on a Krasner hyperfield have more codes words with the same parameters.

Keywords: Linear codes, Fuzzy set, Krasner hyperstructures, Fuzzy logic, Algebraic hyperstructures

1. Introduction

In mathematics, non classical structures as fuzzy sets and algebraic hyperstructures approach better many well known real life situation, and represent natural extension of classical algebraic structures.

Regarding fuzzy sets theory (fuzzy logic), this was introduced in the middle of 1960 by Lotfi Zadeh [1]. This concept is considered today as one of the most important of the second half of twentieth century, this in view of its applications in technological sciences and the impressive quantities of paper and book related to it.

As for algebraic hyperstructures, they were introduced by a french mathematician F. Marty [2] in 1934. Since then, more than a thousand papers and several book have been written on this topic. A well known type of algebraic hyperstructures is due to Krasner [3], who used as a technical tool in a study of his on the approximation of valued fields. In the literature they are called Krasner hyperrings and Krasner hyperfields.

Transmission on coding theory always suppose to encode its information and decode the received information, this is what the classical coding theory introduced in 1948 by C. Shannon [4] deals with. It should ne noted that the handling information are certains. So how can we do if the informations are uncertain? Thus as a new perspective for the algebraic coding, we present below a connection between fuzzy sets, Krasner hyperstructures and linear codes, and we find out how they can bring more in classical coding theory.

IntechOpen

2. Fuzzy linear codes over $\mathbb{Z}_{p^k}$

2.1 Preliminaries

The theory of fuzzy code as we present here were introduce by Shum and Chen De Gang [5], although they have authors such as Hall Diall and Von Kaenel [6, 7] who also worked on it. In this section, we shall formulate the preliminary definitions and results that are required for a good understanding of the sequel (we can see it in [8–10]).

Definition 2.1. Let X be a non-empty set, let I and J be two fuzzy subsets in X, then:

- $(I \cap J)(x) = min\{I(x), J(x)\}$, $(I \cup J)(x) = max\{I(x), J(x)\}$,
- $I = J$ if and only if $I(x) = J(x)$, $I \subseteq J$ if and only if $I(x) \leq J(x)$,
- $(I + J)(x) = max\{I(y) \wedge J(z) | x = y + z\}$, $(IJ)(x) = max\{I(y) \wedge J(z) | x = yz\}$.

These for all $x, y, z \in X$.

Let denoted by M the $\mathbb{Z}_{p^k}$-module $\mathbb{Z}_{p^k}^n$, where p is a prime integer and $n, k \in \mathbb{N} \backslash \{0\}$.

The following definitions on the fuzzy linear space derive from [11, 12].

Definition 2.2. We called a fuzzy submodule of M, a fuzzy subset $\mathcal{F}\sqcap$ of a $\mathbb{Z}_{p^k}$-module M such that for all $x, y \in M$ and $r \in \mathbb{Z}_{p^k}$, we have:

- $\mathcal{F}(x + y) \geq min\{\mathcal{F}\sqcap(x), \mathcal{F}\sqcap(y)\}$.
- $\mathcal{F}(rx) \geq \mathcal{F}\sqcap(x)$.

Definition 2.3. Let $\mathcal{F}\sqcap$ be a fuzzy subset of a nonempty set M. For $t \in [0, 1]$, we called the the upper t-level cut and lower t-level cut of $\mathcal{F}\sqcap$, the sets $\mathcal{F}\sqcap_t = \{x \in M | \mathcal{F}\sqcap(x) \geq t\}$ and $\overline{\mathcal{F}\sqcap_t} = \{x \in M | \mathcal{F}\sqcap(x) \leq t\}$ respectively.

Proposition 2.4. *$\mathcal{F}\sqcap$ is a fuzzy submodule of an $\mathbb{Z}_{p^k}$-module M if and only if for all $\alpha, \beta \in \mathbb{Z}_{p^k}$; $x, y \in M$, we have $\mathcal{F}\sqcap(\alpha x + \beta y) \geq min\{\mathcal{F}\sqcap(x), \mathcal{F}\sqcap(y)\}$.*

The following difinition recalled the notion on fuzzy ideal of a ring.

Definition 2.5. *We called a fuzzy ideal of $\mathbb{Z}_{p^k}$, a fuzzy subset I of a ring $\mathbb{Z}_{p^k}$ such that for each $x, y \in \mathbb{Z}_{p^k}$;*

- $I(x - y) \geq min\{I(x), I(y)\}$.
- $I(xy) \geq max\{I(x), I(y)\}$.

Definition 2.6. Let G be a group and R a ring. We denote by **RG** the set of all formal linear combinations of the form $\alpha = \sum_{g \in G} a_g g$ (where $a_g \in R$ and $a_g = 0$ almost everywhere, that is only a finite number of coefficients are different from zero in each of these sums).

Definition 2.7. Let **RG** a ring group which is the group algebra of $<x>$ on the ring $\mathbb{Z}_{p^k}$ (where x is an invertible element of $\mathbb{Z}_{p^k}$). A fuzzy subset I of **RG** is called a fuzzy ideal of **RG**, if for all $\alpha, \beta \in$ **RG**,

- $I(\alpha\beta) \geq max\{I(\alpha), I(\beta)\}$.

- $I(\alpha - \beta) \geq min\ \{I(\alpha), I(\beta)\}$.

When we use the transfer principle in [13], we easily get the next Proposition.

Proposition 2.8. *A is a fuzzy ideal of RG if and only if for all* $t \in [0, 1]$, *if* $A_t \neq \emptyset$, *then* A_t *is an ideal of RG.*

The following is very important, the give the meaning of the linear code over the ring $\mathbb{Z}_{p^k}$.

Definition 2.9. A submodule of $\mathbb{Z}_{p^k}^n$, is called a linear code of length n over $\mathbb{Z}_{p^k}$. (with n a positive integer).

Contrary to the vector spaces, the module do not admit in general a basis. However it possesses a generating family and therefore a generating matrix, but the decomposition of the elements on this family is not necessarily unique.

Definition 2.10. We called generating matrix of a linear code over $\mathbb{Z}_{p^k}$ all matrix of $\mathcal{M}\left(\mathbb{Z}_{p^k}\right)$, where the lines are the minimal generating family of code.

The equivalence of two codes is define by the following definition.

Definition 2.11. Let C_{p^k} and C'_{p^k} two linear codes over $\mathbb{Z}_{p^k}$ of generating matrix G and G' respectively. The codes C_{p^k} and C'_{p^k} are equivalences if there exists a permutation matrix P, such that $G' = GP$.

To define a dual of a code which is helpful when we fine some properties of the codes, we need to know the inner product.

Definition 2.12. Let C_{p^k} be a linear code of length n over $\mathbb{Z}_{p^k}$, the dual of the code C_{p^k} that we note $C_{p^k}^{\perp}$ is the submodule of $\mathbb{Z}_{p^k}^n$ define by; $C_{p^k}^{\perp} = \{a|$ for all $b \in C_{p^k}, a.b = 0\}$. where "$\cdot$" is the natural inner product on the submodule $\mathbb{Z}_{p^k}^n$.

In a linear code C_{p^k} of length n over $\mathbb{Z}_{p^k}$, if for all $(a_0, \cdots, a_{n-1}) \in C_{p^k}$, then $s((a_0, \cdots, a_{n-1})) \in C_{p^k}$ (where s is the shift map), then the code is said to cyclic.

2.2 On fuzzy linear codes over $\mathbb{Z}_{p^k}$

Now we bring fuzzy logic in linear codes and introduce the notion of fuzzy linear code over the ring $\mathbb{Z}_{p^k}$.

Definition 2.13. Let $M = \mathbb{Z}_{p^k}^n$ be a $\mathbb{Z}_{p^k}$-module. The fuzzy submodule $\mathcal{F}\sqcap$ of M is called fuzzy linear code of length n over $\mathbb{Z}_{p^k}$.

Using the transfer principle of Kondo [13], we have what is follow.

Proposition 2.14. *Let A be a fuzzy set on* $\mathbb{Z}_{p^k}^n$.

A is a fuzzy linear code of length n over $\mathbb{Z}_{p^k}$ if and only if for any $t \in [0, 1]$, if $A_t \neq \emptyset$, then A_t is a linear code of length n over $\mathbb{Z}_{p^k}$.

Corollary 2.15. *Let A be a fuzzy set on* $\mathbb{Z}_{p^k}^n$.

A is a fuzzy linear code of length n over $\mathbb{Z}_{p^k}$ if and only if the characteristic function of any upper t-level cut $A_t \neq \emptyset$ for $t \in [0, 1]$ is a fuzzy linear code of length n over $\mathbb{Z}_{p^k}$.

Example 2.16. *Consider a fuzzy subset* $\mathcal{F}\sqcap$ *on* $\mathbb{Z}_4$ *as follows:*

$$\mathcal{F}\sqcap : \mathbb{Z}_4 \to [0, 1], x \mapsto \begin{cases} 1 & if \ \ x = 0; \\ \frac{1}{3} & if \ \ x = 1; \\ \frac{1}{3} & if \ \ x = 2; \\ \frac{1}{3} & if \ \ x = 3. \end{cases}.$$

Then $\mathcal{F}\sqcap$ is a fuzzy submodule on $\mathbb{Z}_4$-module $\mathbb{Z}_4$, hence $\mathcal{F}\sqcap$ is a fuzzy linear code over $\mathbb{Z}_4$.

Remark 2.17. Let $\mathcal{F}\sqcap$ be a fuzzy linear code of length n over $\mathbb{Z}_{p^k}$, since $\mathbb{Z}_{p^k}^n$ is a finite set, then $Im(\mathcal{F}\sqcap) = \left\{\mathcal{F}\sqcap(x)|x \in \mathbb{Z}_{p^k}^n\right\}$ is finite. Let $Im(\mathcal{F}\sqcap)$ is set as: $t_1 > t_2 > \cdots > t_m$ (where $t_i \in [0,1]$) that is $Im(\mathcal{F}\sqcap)$ have m elements.

Since $\mathcal{F}\sqcap_{t_i}$ is a linear code over $\mathbb{Z}_{p^k}$, let G_{t_i} his generator matrix, $\mathcal{F}\sqcap$ can be determined by m matrixes G_{t_1}, G_{t_2}, $\cdots$, G_{t_m} as in the below Theorem 2.31.

There are some know notions of the orthogonality in fuzzy space, but no one of them does not hold here because these definitions does not meet the transfer principle in the sense of the orthogonality for the t-level cut sets. So we have to introduce an new notion of orthogonality on fuzzy submodules.

Definition 2.18. Let $\mathcal{F}\sqcap_1$ and Fu_2 be two fuzzy submodules on module $\mathbb{Z}_{p^k}^n$ over the ring $\mathbb{Z}_{p^k}$. We said that $\mathcal{F}\sqcap_1$ and $\mathcal{F}\sqcap_2$ are orthogonal if $Im(\mathcal{F}\sqcap_2) = \{1-\alpha|\ \alpha \in Im(\mathcal{F}\sqcap_1)\}$ and for all $t \in [0,1]$, $(\mathcal{F}\sqcap_2)_{1-t} = \left((\mathcal{F}\sqcap_1)_t\right)^{\perp} = \{y \in \mathbb{Z}_{p^k}^n | <x,y> = 0,\ \text{for all}\ x \in (\mathcal{F}\sqcap_1)_t\}$. Where $<,>$ is the standard inner product on $\mathbb{Z}_{p^k}^n$.

Noted that $\mathcal{F}\sqcap_1 \perp \mathcal{F}\sqcap_2$ means $\mathcal{F}\sqcap_1$ and $\mathcal{F}\sqcap_2$ are orthogonal. We what is follow as an example.

Example 2.19. *Consider the two fuzzy submodules $\mathcal{F}\sqcap_1$ and $\mathcal{F}\sqcap_2$ on $\mathbb{Z}_4$ defined as follows:*

$$\mathcal{F}\sqcap_1 : \mathbb{Z}_4 \to [0,1], x \mapsto \begin{cases} \frac{1}{2} & \text{if } x = 0; \\ \frac{1}{4} & \text{if } x = 1; \\ \frac{1}{3} & \text{if } x = 2; \\ \frac{1}{4} & \text{if } x = 3. \end{cases} \quad \text{and } \mathcal{F}\sqcap_2 : \mathbb{Z}_4 \to [0,1],$$

$$x \mapsto \begin{cases} \frac{3}{4} & \text{if } x = 0; \\ \frac{1}{2} & \text{if } x = 1; \\ \frac{2}{3} & \text{if } x = 2; \\ \frac{1}{2} & \text{if } x = 3. \end{cases}.$$

We easily observe that:

$(\mathcal{F}\sqcap_1)_{\frac{1}{2}} = \{0\}$ and $(\mathcal{F}\sqcap_2)_{\frac{1}{2}} = \mathbb{Z}_4$,

$(\mathcal{F}\sqcap_1)_{\frac{1}{4}} = \mathbb{Z}_4$ and $(\mathcal{F}\sqcap_1)_{\frac{3}{4}} = \{0\}$,

$(\mathcal{F}\sqcap_1)_{\frac{1}{3}} = \{0,2\}$ and $(\mathcal{F}\sqcap_1)_{\frac{2}{3}} = \{0,2\}$.

Thus $\mathcal{F}\sqcap_1 \perp \mathcal{F}\sqcap_2$.

Remark 2.20. Let $\mathcal{F}\sqcap_1$ be a fuzzy submodule on module $\mathbb{Z}_{p^k}^n$ such that $\forall x \in \mathbb{Z}_{p^k}^n$, $\mathcal{F}\sqcap_1(x) = \gamma$ (with $\gamma \in [0,1]$), then it does not exists a fuzzy set $\mathcal{F}\sqcap$ on $\mathbb{Z}_{p^k}^n$ such that $\mathcal{F}\sqcap_1 \perp \mathcal{F}\sqcap$.

The previous Remark 2.20 show that the orthogonal of some fuzzy submodule in our sense does not always exists, so it is important to see under which condition the orthogonal of fuzzy submodule exists. The following theorem show the existence of the orthogonal of some fuzzy submodule.

Theorem 2.21. *Let $\mathcal{F}\sqcap_1$ be a fuzzy submodule on a finite module $\mathbb{Z}_{p^k}^n$. If $Im(\mathcal{F}\sqcap_1)$ have more that one element and for all $\varsigma \in Im(\mathcal{F}\sqcap_1)$ there exist $\epsilon \in Im(\mathcal{F}\sqcap_1)$ such that $A_\varsigma = (A_\epsilon)^\perp$, then there always exists a fuzzy submodule $\mathcal{F}\sqcap_2$ on $\mathbb{Z}_{p^k}^n$ such that $\mathcal{F}\sqcap_1 \perp \mathcal{F}\sqcap_2$.*

Proof. Let $\mathcal{F}\sqcap_1$ be a fuzzy submodule on $\mathbb{Z}_{p^k}^n$. Assume that $|Im(\mathcal{F}\sqcap_1)| = m > 1$ and for any $\varsigma \in Im(A)$ there exist $\epsilon \in Im(\mathcal{F}\sqcap_1)$ such that $(\mathcal{F}\sqcap_1)_\varsigma = ((\mathcal{F}\sqcap_1)_\epsilon)^\perp$.

Assume that $Im(\mathcal{F}\sqcap_1) = \{t_1 > t_2 > \cdots > t_m\}$. Let the sets $M_i = \left\{x \in \mathbb{Z}_{p^k}^n | \mathcal{F}\sqcap_1(x) = t_i\right\}$, $i = 1, \cdots, m$. These sets form a partition of $\mathbb{Z}_{p^k}^n$.

Let us define a fuzzy set $\mathcal{F}\sqcap$ as follow:

$\mathcal{F}\sqcap : \mathbb{Z}_{p^k}^n \to [0, 1], x \mapsto 1 - t_{m-i+1}$, if $x \in M_i$.

Since $Im(\mathcal{F}\sqcap_1) = \{t_1 > t_2 > \cdots > t_m\}$, we have $(\mathcal{F}\sqcap_1)_{t_1} \subseteq (\mathcal{F}1)_{t_2} \subseteq \cdots \subseteq (\mathcal{F}\sqcap_1)_{t_m}$. As for any $\varsigma \in Im(\mathcal{F}\sqcap_1)$ there exist $\epsilon \in Im(A)$ such that $A_\varsigma = (A_\epsilon)^\perp$, then $A_{t_i} = \left(A_{t_{m-i+1}}\right)^\perp$.

Thus $\mathcal{F}\sqcap_{1-t_{m-i+1}} = \left\{x \in \mathbb{Z}_{p^k}^n | \mathcal{F}\sqcap(x) \geq 1 - t_{m-i+1}\right\} = M_i \cup M_{i-1} \cup \cdots \cup M_1 = (\mathcal{F}\sqcap_1)_{t_i} = \left((\mathcal{F}\sqcap_1)_{t_{m-i+1}}\right)^\perp$.

Then by taken $\mathcal{F}\sqcap_2 = \mathcal{F}\sqcap$ we obtain the need fuzzy submodule. □

When the orthogonality exist, there is unique. We have the following theorem to show it.

Theorem 2.22. *Let $\mathcal{F}\sqcap_1$, $\mathcal{F}\sqcap_2$ and $\mathcal{F}\sqcap_3$ be three fuzzy submodules on module $\mathbb{Z}_{p^k}^n$, such that $\mathcal{F}\sqcap_1 \perp \mathcal{F}\sqcap_2$ and $\mathcal{F}\sqcap_1 \perp \mathcal{F}\sqcap_3$, then $\mathcal{F}\sqcap_2 = \mathcal{F}\sqcap_3$.*

Proof. Consider that $\mathcal{F}\sqcap_1 \perp \mathcal{F}\sqcap_2$ and $\mathcal{F}\sqcap_1 \perp \mathcal{F}\sqcap_1 3$.

Let $t \in [0, 1]$, and $b \in (\mathcal{F}\sqcap_2)_{1-t}$, then $<a, b> = 0$, for all $a \in (\mathcal{F}\sqcap_1)_t$. Thus $b \in (\mathcal{F}\sqcap_3)_{1-t}$ and $(\mathcal{F}\sqcap_2)_{1-t} \subseteq (\mathcal{F}\sqcap_3)_{1-t}$. Therefore $(\mathcal{F}\sqcap_3)_t \subseteq (\mathcal{F}\sqcap_3)_t$.

In the same way, we show that $(\mathcal{F}\sqcap_2)_t \subseteq (\mathcal{F}\sqcap_3)_t$. Therefore $\mathcal{F}\sqcap_2 = \mathcal{F}\sqcap_3$. □

Corollary 2.23. *The orthogonal of a fuzzy set on $\mathbb{Z}_{p^k}^n$ is a fuzzy submodule on $\mathbb{Z}_{p^k}^n$.*

The orthogonality is an indempotent operator, in fact if $\mathcal{F}\sqcap$ be a fuzzy submodule on $\mathbb{Z}_{p^k}^n$ then $(\mathcal{F}\sqcap^\perp)^\perp = \mathcal{F}\sqcap_1$.

The notion of equivalence on fuzzy linear code can be define as follow.

Definition 2.24. Let $\mathcal{F}\sqcap_1$ and $\mathcal{F}\sqcap_2$ be two fuzzy linear codes over $\mathbb{Z}_{p^k}$. $\mathcal{F}\sqcap_1$ and $\mathcal{F}\sqcap_2$ are said to be equivalent if for all $t \in [0, 1]$, the linear codes $(\mathcal{F}\sqcap_1)_t$ and $(\mathcal{F}\sqcap_2)_t$ are equivalent.

Example 2.25. *Let C_{G_1} and C_{G_2} be two equivalent linear codes of length n over $\mathbb{Z}_{p^k}$.*

We define the equivalent fuzzy linear codes as follow:

$$\mathcal{F}\sqcap_1 : \mathbb{Z}_{p^k}^n \to [0, 1], x \mapsto \begin{cases} 1 & \textit{if } x \in C_{G_1}; \\ 0 & \textit{otherwise}. \end{cases} \text{ and.}$$

$$\mathcal{F}\sqcap_2 : \mathbb{Z}_{p^k}^n \to [0, 1], x \mapsto \begin{cases} 1 & \textit{if } x \in C_{G_2}; \\ 0 & \textit{otherwise}. \end{cases}.$$

Thus the 1 and 0 -level cut of the both fuzzy linear codes give $(\mathcal{F}\sqcap_1)_1 = C_{G_1}$ and $(\mathcal{F}\sqcap_2)_1 = C_{G_2}$,

$(\mathcal{F}\sqcap_1)_0 = \mathbb{Z}_{p^k}^n$ and $(\mathcal{F}\sqcap_2)_0 = \mathbb{Z}_{p^k}^n$.

Remark 2.26. Two equivalent fuzzy linear codes over $\mathbb{Z}_{p^k}$ have the same image.

2.3 Fuzzy linear codes in a practical way

As we have said in the introduction, how fuzzy linear code can deal with uncertain information in a practical way? This subsection allow us to use directly fuzzyness in the information theory.

Let us draw the communication channel as follows:

$$F^k \xrightarrow{Encoding} F^n \xrightarrow{Channel} \mathbb{R}^n \xrightarrow{Decoding} F^k$$

Assume that $R^k = \mathbb{Z}_2^2$ and $R^n = \mathbb{Z}_2^3$, that means that $k = 2$ and $n = 3$. Let $C \subseteq R^3$ be a linear code over R, in the classical case, when we send a codeword $a = (101) \in C$ through a communication channel, the signal receive can be read as $a' = (0.98, 0.03, 0.49)$ and modulate to $a'' = (100)$. Thus to know if a'' belong to the code C, we use syndrome calculation [14]. Since the modulation have gave a wrong word, we can consider that a' have more information than a'', in the sense that we can estimate a level to which a word 0 is modulate to 1, and a word 1 is modulate to 0. Therefore it is possible to use the idea of fuzzy logic to recover the transmit codeword.

Let C be a linear code over $\mathbb{Z}_2^3$. To each $a \in C$, we find $t \in [0, 1]$ such that t estimate the degree of which the element of $\mathbb{R}^3$, obtain from a through the transmission channel belong to the code C. Thus in $\mathbb{Z}_2^3$ the information that we handle are certain, whereas in $\mathbb{R}^3$ there are uncertain. When we associate to all elements of $\mathbb{Z}_2^3$ the degree of which its correspond element obtain through the transmission channel belong to $\mathbb{Z}_2^3$, then we obtain a fuzzy code. If the fuzzy code are fuzzy linear code, then we can recover the code C just by using the upper t-level cut. Thus we deal directly with the uncertain information to obtain the code C.

The following example illustrate this reconstruction of the code by using uncertain information in the case of fuzzy linear code.

Example 2.27. *Let* $\mathbb{Z}_2^3 = \{000, 001, 010, 100, 110, 101, 011, 111\}$ *and* $C = \{000, 001, 110, 111\}$ *be a linear code over* $\mathbb{Z}_2$.

Assume that after the transmission we obtain respectively $\{000; 0.01, 01; 1.01, 10; 1.001, 1, 0.999\}$. Let $\mathcal{F}\sqcap : \mathbb{Z}_2^3 \to [0, 1]$ such that

$$x \mapsto \begin{cases} \{1\} & \text{if } x = 000; \\ \{0.99\} & \text{if } x = 001; \\ \{0.9\} & \text{if } x = 010; \\ \{0.9\} & \text{if } x = 100; \\ \{0.99\} & \text{if } x = 110; \\ \{0.9\} & \text{if } x = 101; \\ \{0.9\} & \text{if } x = 011; \\ \{0.99\} & \text{if } x = 111. \end{cases}.$$

Then by finding a $t \in [0, 1]$ such that $\mathcal{F}\sqcap_t = \{x \in \mathbb{Z}_2^3 | \ \mathcal{F}\sqcap(x) \geq t\} = C$, we obtain $t > 0.9$. Thus, for $t = 0.99$, we are sure that the receive codeword is in C.

It should be better to investigate in deep this approach.

2.4 Fuzzy cyclic code over $\mathbb{Z}_{p^k}$

Let the module $\mathbb{Z}_{p^k}^n$, in this subsection we will consider the case where the integers n and p are coprime.

Definition 2.28. A fuzzy module $\mathcal{F}\sqcap$ on the module $\mathbb{Z}_{p^k}^n$ is called a fuzzy cyclic code of length n over $\mathbb{Z}_{p^k}$ if for all $(a_0, a_1, \cdots, a_{n-1}) \in \mathbb{Z}_{p^k}^n$, then $\mathcal{F}\sqcap((a_{n-1}, a_0, \cdots, a_{n-2})) \geq \mathcal{F}\sqcap((a_0, a_1, \cdots, a_{n-1}))$.

The following proposition give a caracterization of the fuzzy cyclic codes.

Proposition 2.29. *[15] A fuzzy submodule* $\mathcal{F}\sqcap$ *on on* $\mathbb{Z}_{p^k}^n$ *is a fuzzy cyclic code if and only if for all.*

$(a_0, a_1, \cdots, a_{n-1}) \in \mathbb{Z}_{p^k}^n$, we have $\mathcal{F}\sqcap((a_0, a_1, \cdots, a_{n-1})) = \mathcal{F}\sqcap((a_{n-1}, a_0, \cdots, a_{n-2})) = \cdots =$

$$\mathcal{F}\sqcap((a_1, a_2, \cdots, a_{n-1}, a_0)).$$

Proposition 2.30. *$\mathcal{F}\sqcap$ is a fuzzy cyclic code of length n over $\mathbb{Z}_{p^k}$ if and only if for all $t \in [0,1]$, if $(\mathcal{F}\sqcap)_t \neq \varnothing$, then $(\mathcal{F}\sqcap)_t$ is a ideal of the factor ring $\frac{\mathbb{Z}_{p^k}[X]}{(X^n-1)}$.*

Proof. Assume that $\mathcal{F}\sqcap$ is a fuzzy cyclic code over $\mathbb{Z}_{p^k}$ and $t \in [0,1]$ such that $(\mathcal{F}\sqcap)_t \neq \varnothing$. Then $(\mathcal{F}\sqcap)_t$ is a cyclic code over $\mathbb{Z}_{p^k}$.

Let $\psi : \mathbb{Z}_{p^k}^n \to \frac{\mathbb{Z}_{p^k}[X]}{(X^n-1)}, \underline{c} = (c_0, \cdots, c_{n-1}) \mapsto \psi(\underline{c}) = \sum_{i=0}^{n-1} c_i X^i$.

It is prove by easy way that ψ is a isomorphism of $\mathbb{Z}_{p^k}$-module, which send a cyclic codes over $\mathbb{Z}_{p^k}$ onto the ideals of the factor ring $\frac{\mathbb{Z}_{p^k}[X]}{(X^n-1)}$. Therefore, $\forall t \in [0,1]$, $\mathcal{F}\sqcap_t$ is a ideal of $\frac{\mathbb{Z}_{p^k}[X]}{(X^n-1)}$.

Conversely, assume that, $\forall t \in [0,1]$ such that $\mathcal{F}\sqcap_t \neq \varnothing$, $\mathcal{F}\sqcap_t$ is a ideal of factor ring $\frac{\mathbb{Z}_{p^k}[X]}{(X^n-1)}$. Since $\mathcal{F}\sqcap_t$ is a ideal of factor ring $\frac{\mathbb{Z}_{p^k}[X]}{(X^n-1)}$, then $\mathcal{F}\sqcap_t$ is a submodule of $\mathbb{Z}_{p^k}$-module $\mathbb{Z}_{p^k}^n$. Hence $\mathcal{F}\sqcap_t \neq \varnothing$, is a linear code over $\mathbb{Z}_{p^k}$, then $\mathcal{F}\sqcap$ is a fuzzy linear code. Due to ψ, $\forall t \in [0,1]$, $\mathcal{F}\sqcap_t$ is a cyclic code over $\mathbb{Z}_{p^k}$, then $\mathcal{F}\sqcap$ is a fuzzy cyclic code over $\mathbb{Z}_{p^k}$. □

Since $\mathbb{Z}_{p^k}$ is a finite ring, then $Im(\mathcal{F}\sqcap) = \left\{\mathcal{F}\sqcap(x) \in [0,1] | x \in \mathbb{Z}_{p^k}^n\right\}$ is also finite. Assume that $Im(\mathcal{F}\sqcap) = \{t_1 > t_2 > \cdots > t_m\}$, then $\mathcal{F}\sqcap_{t_1} \subseteq \mathcal{F}\sqcap_{t_2} \subseteq \cdots \subseteq \mathcal{F}\sqcap_{t_{m-1}} \subseteq A_{t_m} = \mathbb{Z}_{p^k}^n$.

Let $g_i^{(k)}(X) \in \mathbb{Z}_{p^k}[X]$ the generator polynomial of $\mathcal{F}\sqcap_{t_i}$, note that $g_i^{(k)}(X)$ is the Hensel lifting of order k of some polynomial $g_i(X) \in \mathbb{Z}_p[X]$ which divide $X^n - 1$, the cyclic code $< g_i^{(k)}(X) > \subset \frac{\mathbb{Z}_{p^k}[X]}{(X^n-1)}$ is called the **lifted code** of the cyclic code $< g_i(X) > \subset \frac{\mathbb{Z}_p[X]}{(X^n-1)}$ [8].

Since $\mathcal{F}\sqcap_{t_1} \subseteq \mathcal{F}\sqcap_{t_2} \subseteq \cdots \subseteq \mathcal{F}\sqcap_{t_{m-1}} \subseteq \mathcal{F}\sqcap_{t_m} = \mathbb{Z}_{p^k}^n$, then $g_{i+1}^{(k)}(X) | g_i^{(k)}(X), i = 1, \cdots, m-1$. So we define the polynomial $h_i^{(k)}(X) = (X^n - 1)/g_i^{(k)}(X)$ which is called the check polynomial of the cyclic code $\mathcal{F}\sqcap_{t_i} = < g_i^{(k)}(X) >, i = 1, \cdots, m$.

Theorem 2.31. *Let $\mathcal{G} = \left\{g_1^{(k)}(X), g_2^{(k)}(X), \cdots, g_m^{(k)}(X)\right\}$ be a set of polynomial in $\mathbb{Z}_{p^k}[X]$, such that $g_i(X)$ divide $X^n - 1$, $i = 1, \cdots, m$. If $g_{i+1}^{(k)}(X) | g_i^{(k)}(X)$ for $i = 1, 2, \cdots, m-1$ and $< g_m^{(k)}(X) > = \mathbb{Z}_{p^k}^n$, then the set $\mathcal{G}$ can determined a fuzzy cyclic code $\mathcal{F}\sqcap$ and $\left\{ < g_i^{(k)}(X) > | i = 1, \cdots, m\right\}$ is the family of upper level cut cyclic subcodes of $\mathcal{F}\sqcap$.*

The proof is leave for the reader but he can check it in [15].

3. Fuzzy $\mathbb{Z}_{p^k}$-linear code

In the previous section, we study define and fuzzy linear codes over the ring $\mathbb{Z}_{p^k}$ in the previous section. Now define the notion on **fuzzy Gray map**, we are going to use it in the construction of the fuzzy $\mathbb{Z}_{p^k}$-linear codes which is different from the fuzzy linear codes over the ring $\mathbb{Z}_{p^k}$.

3.1 Fuzzy Gray map

When we order and enumerate a binary sequences of a fixed length we obtain the code of Gray in it original form. For the length two which interest us directly we have the following Gray code:

$$0 \mapsto 00$$

$$1 \mapsto 01$$

$$2 \mapsto 11$$

$$3 \mapsto 10.$$

Let $\phi : \mathbb{Z}_{2^2} \to \mathbb{Z}_2^2$ the Gray map.

Using the extension principle [16], we will define the fuzzy Gray map between two fuzzy spaces by what is follow.

Definition 3.1. Consider the Gray map $\phi : \mathbb{Z}_{2^2} \to \mathbb{Z}_2^2$. Let $\mathcal{F}(\mathbb{Z}_{2^2})$, $\mathcal{F}(\mathbb{Z}_2^2)$ the set of all the fuzzy subset on $\mathbb{Z}_{2^2}$ and $\mathbb{Z}_2^2$ respectively. The fuzzy Gray map is the map $\hat{\phi} : \mathcal{F}(\mathbb{Z}_{2^2}) \to \mathcal{F}(\mathbb{Z}_2^2)$, such that for all $\mathcal{F}\sqcap \in \mathcal{F}(\mathbb{Z}_{2^2})$, $\hat{\phi}(\mathcal{F}\sqcap)(y) = sup\{A(x) | y = \phi(x)\}$.

The next Theorem is straightforward.

Theorem 3.2. *The fuzzy Gray map* $\hat{\psi}$ *is a bijection.*

Proof: It is due to the fact that ψ is one to one function. □

As in crisp case, we have the following Proposition which is very important.

Proposition 3.3. *If* $\mathcal{F}\sqcap$ *is a fuzzy linear code over* $\mathbb{Z}_{2^2}$ *and* ϕ *the Gray map, then* $\hat{\phi}(\mathcal{F}\sqcap)$ *is no always a fuzzy linear code over the field* $\mathbb{Z}_2$.

The Gray map give a way to construct the nonlinear codes as binary image of the linear codes, we have for example the case of Kerdock, Preparata, and Goethals codes which have very good properties and also useful (We refer reader for it on [17, 18]). Moreover if $\mathcal{C}$ is a linear code of length n over $\mathbb{Z}_4$, then $C = \psi(\mathcal{C})$ is a nonlinear code of length $2n$ over $\mathbb{Z}_2$ in generally [18]. In that way we construct a fuzzy Kerdock code in the following example.

Example 3.4. *Let* $G = \begin{pmatrix} 1 & 0 & 0 & 0 & 2 & 1 & 1 & 1 \\ 0 & 1 & 0 & 0 & 1 & 2 & 1 & 3 \\ 0 & 0 & 1 & 0 & 1 & 3 & 2 & 1 \\ 0 & 0 & 0 & 1 & 1 & 1 & 3 & 2 \end{pmatrix}$ *be a generating matrix for a linear code* $\mathcal{C}$ *of length* 8 *over* $\mathbb{Z}_4$. *Then his image under the Gray map* ϕ *give a Kerdock code* C.

Let $\mathcal{F}\sqcap : \mathbb{Z}_4^8 \to [0,1], x \mapsto \begin{cases} 1, & if \ x \in \mathcal{C}; \\ 0, & otherwise. \end{cases}$ Thus $\mathcal{F}\sqcap$ is a fuzzy linear code over $\mathbb{Z}_4$.

Since ϕ is a bijection, we construct $\hat{\phi}(\mathcal{F}\sqcap) : \mathbb{Z}_2^{16} \to [0,1], y \mapsto \begin{cases} 1, & y \in \mathcal{E}; \\ 0, & otherwise. \end{cases}$,

where $\mathcal{E} = \{y \in \mathbb{Z}_2^{16} | y = \phi(x) \text{ and } x \in \mathcal{C}\}$.

Noted that as $\mathcal{E}$ is not a linear code over $\mathbb{Z}_2$, then $\hat{\phi}(Fu)$ is a fuzzy $\mathbb{Z}_2$-linear code but not a fuzzy linear code over $\mathbb{Z}_2$.

$\hat{\phi}(\mathcal{F}\sqcap)$ is a fuzzy Kerdock code of length 16.

By the Example, we remark that a fuzzy $\mathbb{Z}_4$-linear code is not in generally a fuzzy linear code over $\mathbb{Z}_2$.

If we define the fuzzy binary relation R_ϕ on $\mathbb{Z}_{2^2} \times \mathbb{Z}_2^2$ by $R_\phi(x,y) = \begin{cases} 1, & \text{if } y = \psi(x); \\ 0, & \text{otherwise.} \end{cases}$ It is easy to see [19] that $\hat{\phi}(\mathcal{F}\sqcap)(y) = sup\{\mathcal{F}\sqcap(x)|y = \phi(x)\}$ can be represented by $\hat{\phi}(\mathcal{F}\sqcap)(y) = sup\{\, min\,\{\mathcal{F}\sqcap(x), R_\phi(x,y)\}|x \in \mathbb{Z}_2^2\}$.

We now define fuzzy generalized gray map. First we consider the generalized Gray map as in [8] $\Phi : \mathbb{Z}_{p^k} \to \mathbb{Z}_p^{p^{k-1}}$.

Definition 3.5. The map $\hat{\Phi} : \mathcal{F}\left(\mathbb{Z}_{p^k}\right) \to \mathcal{F}\left(\mathbb{Z}_p^{p^{k-1}}\right)$, such that for any $\mathcal{F}\sqcap \in \mathcal{F}\left(\mathbb{Z}_{p^k}\right)$,

$$\hat{\Phi}(\mathcal{F}\sqcap)(y) = \begin{cases} sup\{\mathcal{F}\sqcap(x)|y = \Phi(x)\}, & \text{if a such x exists;} \\ 0, & \text{otherwise.} \end{cases}$$

Is called a fuzzy generalized gray map.

Remark 3.6.

1. The Definition 3.5 can be simply write $\hat{\Phi}(A)(y) = \begin{cases} \mathcal{F}\sqcap(x), & \text{if } y = \Phi(x); \\ 0, & \text{otherwise.} \end{cases}$
   This because $\Phi : \mathbb{Z}_{p^k} \to \mathbb{Z}_p^{p^{k-1}}$ cannot give more than one image for one element.

2. Let $\mathcal{F}\sqcap_1 \in \mathcal{F}\left(\mathbb{Z}_p^{p^{k-1}}\right)$ such that $\mathcal{F}\sqcap_1(y) = t \neq 0$ for any $y \in \mathbb{Z}_p^{p^{k-1}}$. There does not exist a fuzzy subset $\mathcal{F}\sqcap \in \mathcal{F}\left(\mathbb{Z}_{p^k}\right)$ such that $\hat{\Phi}(\mathcal{F}\sqcap) = \mathcal{F}\sqcap_1$.

Thus $\hat{\Phi}$ is not a bijection map.

3.2 Fuzzy $\mathbb{Z}_{p^k}$-linear code

In the following, we will note $\hat{\Phi}$ the map on $\mathcal{F}\left(\mathbb{Z}_{p^k}^n\right)$ onto $\mathcal{F}\left(\mathbb{Z}_p^{n.p^{k-1}}\right)$ which spreads the fuzzy generalized Gray map.

Let define fuzzy $\mathbb{Z}_{p^k}$-linear code.

Definition 3.7. A fuzzy code *Fu* over $\mathbb{Z}_p$ is a fuzzy $\mathbb{Z}_{p^k}$-linear code if it is an image under the fuzzy generalized Gray map of a fuzzy linear code over the ring $\mathbb{Z}_{p^k}$.

For a fuzzy $\mathbb{Z}_{p^k}$-linear code, if it is the image under the generalized Gray map of a cyclic code over the ring $\mathbb{Z}_{p^k}$. Then the fuzzy code *Fu* is called a fuzzy $\mathbb{Z}_{p^k}$-cyclic code.

Remark 3.8. A fuzzy $\mathbb{Z}_{p^k}$-linear code is a fuzzy code over the fields $\mathbb{Z}_p$.

Example 3.9.

$$\text{Let } \mathcal{F}\sqcap : \mathbb{Z}_2^6 \to [0,1], w = (a,b,c,d,e,f) \mapsto \begin{cases} 1, & \textit{if } e = f = 0; \\ 0, & \textit{otherwise.} \end{cases}$$

$\mathcal{F}\sqcap$ is a fuzzy linear code of length 6 over $\mathbb{Z}_2$.

$$\text{Let } \mathcal{F}\sqcap' : \mathbb{Z}_4^3 \to [0,1], v = (x,y,z) \mapsto \begin{cases} 1, & \textit{if } z = 0; \\ 0, & \textit{otherwise.} \end{cases}$$

$\mathcal{F}\sqcap'$ is a fuzzy linear code of length 3 over $\mathbb{Z}_4$.

Since $\mathcal{F}\sqcap = \hat{\phi}(\mathcal{F}\sqcap')$. Then $\mathcal{F}\sqcap'$ is a fuzzy $\mathbb{Z}_4$-linear code.

Using crisp case technic we prove he following Proposition.

Proposition 3.10. *Let Fu be a fuzzy $\mathbb{Z}_{p^k}$-linear code, then Fu is no always a fuzzy linear code over the field $\mathbb{Z}_p$.*

Proof. The need here is to construct an counter-example, which is done in the Example 3.9. □

The following diagram give a construct the fuzzy $\mathbb{Z}_{p^k}$-linear code. This holds because the fuzzy generalized Gray map image of fuzzy linear code can be a fuzzy linear code over the field $\mathbb{Z}_p$:

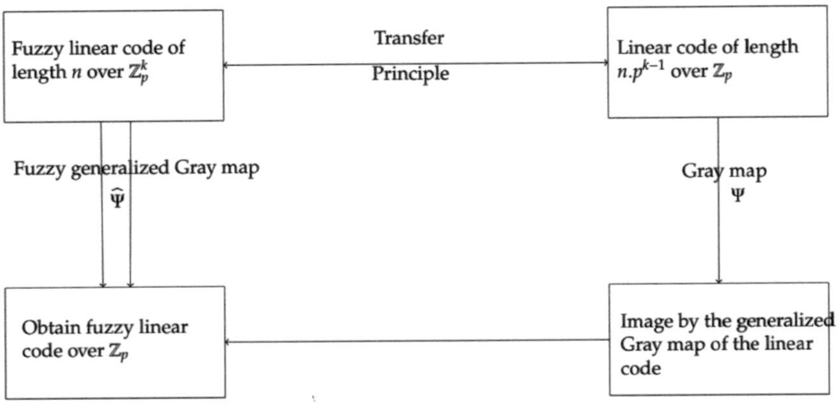

We construct some codes using the both methods.

Example 3.11.

1. *Let $\mathcal{F}\sqcap : \mathbb{Z}_{p^k}^n \to [0,1]$ be a linear code such that $\mathcal{F}\sqcap$ have three upper t-level cut $\mathcal{F}\sqcap_{t_3} \subseteq \mathcal{F}\sqcap_{t_2} \subseteq \mathcal{F}\sqcap_{t_1}$. Let $\mathcal{F}\sqcap'_{t_3} = \Phi(\mathcal{F}\sqcap_{t_3})$, $\mathcal{F}\sqcap'_{t_2} = \Phi(\mathcal{F}\sqcap_{t_2})$ and $\mathcal{F}\sqcap'_{t_1} = \Phi(\mathcal{F}\sqcap_{t_1})$, we have $\mathcal{F}\sqcap'_{t_3} = \Phi(\mathcal{F}\sqcap_{t_3}) \subseteq \mathcal{F}\sqcap'_{t_2} = \Phi(\mathcal{F}\sqcap t_2) \subseteq \mathcal{F}\sqcap'_{t_1} = \Phi(\mathcal{F}\sqcap_{t_1})$. We construct $\mathcal{F}\sqcap' = \hat{\Phi}(\mathcal{F}\sqcap)$ as follow:*

$$\mathcal{F}\sqcap' : \mathbb{Z}_p^{n.p^{k-1}} \to [0,1], y \mapsto \begin{cases} t_3, & \text{if } y \in \mathcal{F}\sqcap'_{t_3}; \\ t_2, & \text{if } y \in \mathcal{F}\sqcap'_{t_2}; \\ t_1, & \text{if } y \in \mathcal{F}\sqcap'_{t_1}; \\ 0, & \text{otherwise.} \end{cases}$$

2. Let $\mathcal{F}\sqcap : \mathbb{Z}_4 \to [0,1], x \mapsto \begin{cases} \frac{1}{2} & \text{if } x = 0; \\ \frac{1}{3} & \text{if } x = 2; \\ \frac{1}{4} & \text{if } x = 1,3. \end{cases}$

be a fuzzy linear code over $\mathbb{Z}_4$. Then $\mathcal{F}\sqcap_{\frac{1}{2}} = \{0\}$, $\mathcal{F}\sqcap_{\frac{1}{3}} = \{0,2\}$ and $\mathcal{F}\sqcap_{\frac{1}{4}} = \mathbb{Z}_4$. We construct $\mathcal{F}\sqcap'_{\frac{1}{2}} = \{00\}$, $\mathcal{F}\sqcap'_{\frac{1}{3}} = \{00, 11\}$ and $\mathcal{F}\sqcap'_{\frac{1}{4}} = \mathbb{Z}_2^2$, the Gray map image of $\mathcal{F}\sqcap_{\frac{1}{2}}$, $\mathcal{F}\sqcap_{\frac{1}{3}}$ and $\mathcal{F}\sqcap_{\frac{1}{4}}$ respectively, we define

$$\mathcal{F}\sqcap' : \mathbb{Z}_2^2 \to [0,1], y \mapsto \begin{cases} \frac{1}{2} & \text{if } x \in \mathcal{F}\sqcap_{\frac{1}{2}},\ y = \phi(x) ; \\ \frac{1}{3} & \text{if } x \in \mathcal{F}\sqcap_{\frac{1}{3}} \setminus \mathcal{F}\sqcap_{\frac{1}{2}},\ y = \phi(x); \\ \frac{1}{4} & \text{if } x \in \mathcal{F}\sqcap_{\frac{1}{4}} \setminus \mathcal{F}\sqcap_{\frac{1}{3}},\ y = \phi(x). \end{cases}$$

We obtain the same code $\mathcal{F}\sqcap'$ and $\hat{\phi}(\mathcal{F}\sqcap)$.

Proposition 3.12. *[15] If for all $t \in [0,1]$, $\mathcal{F}\sqcap'_t = \Phi(\mathcal{F}\sqcap_t)$ (when $\mathcal{F}\sqcap_t \neq \emptyset$) is a linear code over $\mathbb{Z}_p$, then this two constructions of the fuzzy $\mathbb{Z}_p$-linear code above are give the same fuzzy code.*

Proof. This follows directly from the definition of the fuzzy generalized Gray map and the fact that the image under the generalized Gray map of a linear code is not a linear code in general. □

4. Linear codes over Krasner hyperfields

4.1 Preliminaries

This section recall notions and results that are required in the sequel. All of this can also be check on [3, 20–22].

Let $\mathcal{H}$ be a non-empty set and $\mathcal{P}^*(\mathcal{H})$ be the set of all non-empty subsets of $\mathcal{H}$. Then, a map $\circledast : \mathcal{H} \times \mathcal{H} \to \mathcal{P}^*(\mathcal{H})$, where $(h_1, h_2) \mapsto h_1 \circledast h_2 \subseteq \mathcal{H}$ is called a hyperoperation and the couple $(\mathcal{H}, \,, \circledast, \,)$ is called a hypergroupoid.

For all non-empty subsets A and B of $\mathcal{H}$ and $h \in \mathcal{H}$, we define $A \circledast B = \bigcup_{a \in A, b \in B} a \circledast b$, $A \circledast h = A \circledast \{h\}$ and $h \circledast B = \{h\} \circledast B$.

Definition 4.1. A canonical hypergroup $(\mathcal{R}, \oplus)$ is an algebraic structure in which the following axioms hold:

1. For any $x, y, z \in \mathcal{R}$, $x \oplus (y \oplus z) = (x \oplus y) \oplus z$,

2. For any $x, y \in \mathcal{R}$, $x \oplus y = y \oplus x$,

3. There exists an additive identity $0 \in R$ such that $0 \oplus x = \{x\}$ for every $x \in \mathcal{R}$.

4. For every $x \in \mathcal{R}$ there exists a unique element x' (an opposite of x with respect to hyperoperation "$\oplus$") in $\mathcal{R}$ such that $0 \in x \oplus x'$,

5. For any $x, y, z \in \mathcal{R}$, $z \in x \oplus y$ implies $y \in x' \oplus z$ and $x \in z \oplus y'$.

Remark 4.2. Note that, in the classical group $(R, +)$, the concept of opposite of $x \in R$ is the same as inverse.

A canonical hypergroup with a multiplicative operation which satisfies the following conditions is called a Krasner hyperring.

Definition 4.3. An algebraic hyperstructure $(\mathcal{R}, \oplus, \cdot)$, where "$\cdot$" is usual multiplication on $\mathcal{R}$, is called a Krasner hyperring when the following axioms hold:

1. $(\mathcal{R}, \oplus)$ is a canonical hypergroup with 0 as additive identity,

2. $(\mathcal{R}, \cdot)$ is a semigroup having 0 as a bilaterally absorbing element,

3. The multiplication "$\cdot$" is both left and right distributive over the hyperoperation "$\oplus$".

A Krasner hyperring is called commutative (with unit element) if $(\mathcal{R}, \cdot)$ is a commutative semigroup (with unit element) and such is denoted $(R, \oplus, \cdot, 0, 1)$.

Definition 4.4. Let $(R, \oplus, \cdot, 0, 1)$ be a commutative Krasner hyperring with unit such that $(R \backslash \{0\}, \cdot, 1)$ is a group. Then, $(R, \oplus, \cdot, 0, 1)$ is called a Krasner hyperfield.

This Example is from Krasner.

Example 4.5. *[?] Consider a field* $(F, +, \cdot)$ *and a subgroup* G *of* $(F\backslash\{0\}, \cdot)$*. Take* $H = F/G = \{aG|\ a \in F\}$ *with the hyperoperation and the multiplication given by:*

$$\begin{cases} aG \oplus bG = \{\bar{c} = cG|\ \bar{c} \in aG + bG\} \\ aG \cdot bG = abG \end{cases}$$

Then $(H, \oplus, \cdot)$ is a Krasner hyperfield.

We now give an example of a finite hyperfield with two elements 0 and 1, that we name $\mathcal{F}_2$ and which will be used it in the sequel.

Example 4.6. *Let* $\mathcal{F}_2 = \{0, 1\}$ *be the finite set with two elements. Then* $\mathcal{F}_2$ *becomes a Krasner hyperfield with the following hyperoperation "*$\oplus$*" and binary operation "*$\cdot$*".*

$\oplus$	0	1
0	$\{0\}$	$\{1\}$
1	$\{1\}$	$\{0,1\}$

and

$\cdot$	0	1
0	0	0
1	0	1

Let $(\mathcal{R}, \oplus, \cdot)$ be a hyperring, A and B be a non-empty subset of $\mathcal{R}$. Then, A is said to be a subhyperring of $\mathcal{R}$ if $(A, \oplus, \cdot)$ is itself a hyperring. A subhyperring A of a hyperring $\mathcal{R}$ is a left (right) hyperideal of $\mathcal{R}$ if $r \cdot a \in A$ $(a \cdot r \in A)$ for all $r \in \mathcal{R}$, $a \in A$. Also, A is called a hyperideal if A is both a left and a right hyperideal. We define $A \oplus B$ by $A \oplus B = \{x|\ x \in a \oplus b$ for some $a \in A, b \in B\}$ and the product $A \cdot B$ is defined by $A \cdot B = \{x|\ x \in \sum_{i=1}^{n} a_i \cdot b_i,$ with $a_i \in A, b_i \in B, n \in \mathbb{N}^*\}$. If A and B are hyperideals of $\mathcal{R}$, then $A \oplus B$ and $A \cdot B$ are also hyperideals of $\mathcal{R}$.

Definition 4.7. An algebraic structure $(\mathcal{R}, \oplus, \cdot)$ (where $\oplus$ and $\cdot$ are both hyperoperations) is called additive-multiplicative hyperring if the satisfies the following axioms:

1. $(\mathcal{R}, \oplus)$ is a canonical hypergroup with 0 as additive identity,

2. $(\mathcal{R}, \cdot)$ is a semihypergroup having 0 as a bilaterally absorbing element, i.e., $x \cdot 0 = 0 \cdot x = 0$,

3. the hypermultiplication "$\cdot$" is distributive with respect to the hyperoperation "$+$",

4. for all $x, y \in \mathcal{R}$, we have $x \cdot (y') = (x') \cdot y = (x \cdot y)'$.

An additive-multiplicative hyperring $(\mathcal{R}, \oplus, \cdot)$ is said to be commutative if $(\mathcal{R}, \cdot)$ is a commutative semihypergroup. and $(\mathcal{R}, \oplus, \cdot)$ is called a hyperring with multiplicative identity if there exists $e \in \mathcal{R}$ such that $x \cdot e = x = e \cdot x$ for every $x \in \mathcal{R}$.

We close this section with the following definition of the ideal in a additive-multiplicative hyperring.

Definition 4.8 A non-empty subset A of an additive-multiplicative hyperring $\mathcal{R}$ is a left (right) hyperideal if,

1. for all $a, b \in A$, then $a \oplus b' \subseteq A$,

2. for all $a \in A$, $r \in \mathcal{R}$, then $r \cdot a \subseteq A$ $(a \cdot r \subseteq A)$.

4.2 Hypervector spaces over hyperfields

We give some properties related to the hypervector space as it is done by Sanjay Roy and Samanta [23] and all these will allow us to characterize linear codes over a Krasner hyperfield.

From now on, and for the rest of this section, by $\mathcal{F}$ we mean a Krasner hyperfield.

Definition 4.9. [23] Let $\mathcal{F}$ be a Krasner hyperfield. A commutative hypergroup $(\mathcal{V}, \oplus_{\mathcal{V}})$ together with a map $\cdot : \mathcal{F} \times \mathcal{V}\int \to \mathcal{V}\int$, is called a hypervector space over $\mathcal{F}$ if for any $a, b \in \mathcal{F}$ and $x, y \in \mathcal{V}\int$, the following conditions hold:

1. $a \cdot \left(x \oplus_{\mathcal{V}\int} y\right) = a \cdot x \oplus_{\mathcal{V}\int} a \cdot y$ (right distributive law),

2. $\left(a \oplus_{\mathcal{V}\int} b\right) \cdot x = a \cdot x \oplus_{\mathcal{V}\int} b \cdot x$ (left distributive law),

3. $a \cdot (b \cdot x) = (ab) \cdot x$ (associative law),

4. $a \cdot (x') = (a') \cdot x = (a \cdot x)'$,

5. $x = 1 \cdot x$.

Let us give that trivial example of a hypervector space.

Example 4.10. *Let $n \in \mathbb{N}$, $\mathcal{F}^n$ is a hypervector space over $\mathcal{F}$ where the composition of elements are as follows:*

$x \oplus y = \{z \in \mathcal{F}^n; z_i \in x_i \oplus y_i, i = 1 \dots n\}$ and $a \cdot x = (a \cdot x_1, a \cdot x_2, \dots, a \cdot x_n)$ for any $x, y \in \mathcal{F}^n$ and $a \in \mathcal{F}$.

Definition 4.11. [23] Let $(\mathcal{V}\int, \oplus, \cdot, 1)$ be a hypervector space over $\mathcal{F}$. A subset $A \subseteq \mathcal{V}\int$ is called a subhypervector space of $\mathcal{V}\int$ if:

1. $A \neq 0$,

2. for all $x, y \in A$, then $x \oplus y' \subseteq A$,

3. for all $a \in \mathcal{F}$, for all $x \in A$, then $a \cdot x \in A$.

Definition 4.12. [23] Let $\mathcal{S}$ be a subset of a hypervector space $\mathcal{V}\int$ over $\mathcal{F}$. $\mathcal{S}$ is said to be linearly independent if for every $x_1, x_2, \dots, x_n$ in $\mathcal{S}$ and for every $a_1, a_2, \dots, a_n$ in $\mathcal{F}$, $(n \in \mathbb{N}\backslash\{0, 1\})$ such that $0 \in a_1 \cdot x_1 + a_2 \cdot x_2 + \dots + a_n \cdot x_n$ implies that $a_1 = a_2 = \dots = a_n = 0$.

If $\mathcal{S}$ is not linearly independent, then we said that $\mathcal{S}$ is linearly dependent.

If $\mathcal{S}$ is a nonempty subset of $\mathcal{V}\int$, then the smallest subhypervector space of $\mathcal{V}$ containing $\mathcal{S}$ is the set define by

$\langle \mathcal{S} \rangle = \cup \left\{ \sum_{i=1}^{n} a_i \cdot x_i | \, x_i \in \mathcal{S}, a_i \in \mathcal{F}, n \in \mathbb{N}\backslash\{0, 1\} \right\} \cup l(\mathcal{S})$, (where $l(\mathcal{S}) = \{a \cdot x | \, a \in \mathcal{F}, x \in \mathcal{S}\}$).

Definition 4.13. [23] Let $\mathcal{V}\int$ be a hypervector space over $\mathcal{F}$. A vector $x \in \mathcal{V}\int$ is said to be a linear combination of the vectors $x_1, x_2, \dots, x_n \in \mathcal{V}\int$ if there exist $a_1, a_2, \dots, a_n \in \mathcal{F}$ such that $x \in a_1 \cdot x_1 + a_2 \cdot x_2 + \cdots + a_n \cdot x_n$ in the hypervector spaces, the notion of basis exists and he have the following definition.

Definition 4.14. *[23] Let $\mathcal{V}\int$ be a hypervector space over $\mathcal{F}$ and $\mathcal{B}$ be a subset of $\mathcal{V}\int$. The set $\mathcal{B}$ is said to be a basis for $\mathcal{V}\int$ if,*

1. $\mathcal{S}$ is linearly independent,

2. every element of $\mathcal{V}\int$ can be expressed as a finite linear combination of elements from $\mathcal{S}$.

4.3 Polynomial hyperring

We assume that $\mathcal{F}$ is such that for all $a, b \in \mathcal{F}$, $a \cdot (b') = (a') \cdot b = (a \cdot b)'$.

Let denote by $\mathcal{F}[x]$ the set of all polynomials in the variable x over $\mathcal{F}$. Let the polynomials $f(x) = \sum_{i=0}^{n} a_i x^i$ and $g(x) = \sum_{i=0}^{m} b_i x^i$ in $\mathcal{F}[x]$.

Let us define the set $\mathcal{P}^*(\mathcal{F})[x] = \{\sum_{k=0}^{n} A_k x^k;$ where $A_k \in \mathcal{P}^*(\mathcal{F}), n \in \mathbb{N}\}$, the hypersum and hypermultiplication of $f(x)$ and $g(x)$ are defined as follows:

$$\overline{\oplus} : \mathcal{F}[x] \times \mathcal{F}[x] \to \mathcal{P}^*(\mathcal{F})[x] \tag{1}$$

$$(f(x), g(x)) \mapsto (f \overline{\oplus} g)(x) = (a_0 \oplus b_0) + (a_1 \oplus b_1)x + \cdots + (a_M \oplus b_M)x^M, \tag{2}$$

$$where\ M = max\,\{n, m\}. \tag{3}$$

$$\bar{\cdot} : F[x] \times F[x] \to \mathcal{P}^*(F)[x] \tag{4}$$

$$(f(x), g(x)) \mapsto (f \bar{\cdot} g)(x) = \sum_{k=0}^{m+n} \left(\sum_{l+j=k} a_l \cdot b_j \right) x^k, if\ deg(f) \geq 1\ and\ deg(g) \geq 1 \tag{5}$$

The following remark is from Jančic-Rašović [24].

Remark 4.15. The algebraic hyperstructure $(\mathcal{F}[x], \overline{\oplus}, \bar{\cdot})$ is an additive-multiplication hyperring.

4.4 Linear codes and cyclic codes over finite hyperfields

In this section we shall define and discuss about the concept of linear and cyclic codes over the finite Krasner hyperfield $\mathcal{F}_2$ from the **Example 4.6**. Let us recall some basics from code theory. Let $\mathcal{C}$ be a linear code, the Hamming distance $d_H(x, y)$ between two vectors $x, y \in \mathcal{C}$ is defined to be the number of coordinates in which x differs from y. The minimum distance of a code $\mathcal{C}$, denoted by $d(\mathcal{C})$, is $d(\mathcal{C}) = \min\{d_H(x, y) | x, y \in \mathcal{C}$ and $x \neq y\}$. In this case we can also compute for a code word $x \in \mathcal{C}$, the integer $w_H(x)$ which is the number of nonzero coordinates in x also called Hamming weight of x.

We denoted by $k = \dim(\mathcal{C})$ the dimension of $\mathcal{C}$ and the code $\mathcal{C}$ is called an (n, k, d) -code which can be represented by his generator matrix [25].

Let us define linear code over $\mathcal{F}_2$.

Definition 4.16. A subhypervector space of the hypervector space $\mathcal{F}_2^n$ is called a linear code $\mathcal{C}$ of length n over $\mathcal{F}_2$.

The concept of dual code is a very useful in the coding theory. Let us define it on the Krasner hyperfield $\mathcal{F}_2$.

Definition 4.17. Let $\mathcal{C}$ be a linear code of length n ($n \geq 2$) over $\mathcal{F}_2$. The dual of $\mathcal{C}$ is also a linear code defined by $\mathcal{C}^{\perp} := \{y \in \mathcal{F}_2^n | \ 0 \in x \cdot y^t, \forall x \in \mathcal{C}\}$.

The code $\mathcal{C}$ is self-dual if $\mathcal{C} = \mathcal{C}^{\perp}$.

Here is an basic example of a linear code and his dual.

Example 4.18. *Let* $C = \{000, 101, 011, 110, 111\}$ *be a linear code of length* 3 *over* F_2. *It's easy to check that the dual of* C *is defined by* $C^{\perp} = \{000, 111\}$.

As in the classical case, the notion of cyclic code on hyperstructures still works with polynomials. So i that way the polynomial $f(x) = a_0 + a_1x^1 + a_2x^2 + \cdots + a_{n-1}x^{n-1}$ of degree at most $n - 1$ over $\mathcal{F}_2$ may be considered as the sequence $a = (a_0, a_1, a_2, \dots, a_{n-1})$ of length n in F_2^n. In fact, there is a correspondence between $\mathcal{F}_2^n$ and the residue class hyperring $\frac{\mathcal{F}_2[x]}{(x^n-1)}$ [25].

$$\xi : \mathcal{F}_2^n \to \frac{\mathcal{F}_2[x]}{(x^n - 1)}$$

$$c = (c_0, c_1, c_2, \dots, c_{n-1}) \mapsto c_0 + c_1x^1 + c_2x^2 + \cdots + c_{n-1}x^{n-1}.$$

Using Theorem 3.7 in [26], the multiplication of x by any element of $\frac{\mathcal{F}_2[x]}{(x^n-1)}$ is equivalent to applying the shift map s of the Definition?? to the corresponding element of $\mathcal{F}_2^n$, so we use the polynomial to define cyclic code.

We are now going to define a distance relation on linear codes over the finite hyperfield $\mathcal{F}_2$, which will allow us to detect if there is an error in a received word.

Proposition 4.19. *The mapping define by*

$$\lceil_{\mathcal{H}} : \mathcal{F}_2^n \times \mathcal{F}_2^n \to \mathbb{N}$$

$$(x, y) \mapsto \lceil_{\mathcal{H}}(x, y) = card\{i \in \mathbb{N} | \ x_i \neq y_i\}$$

is a distance on $\mathcal{F}_2^n$, called the Hamming distance.

Proof. The proof is similar to the classical case. □

The following remark will be helpful to define Hamming weight.

Remark 4.20. For an $x \in \mathcal{F}_2^n$, we write $x = (\{x_1\}, \dots, \{x_n\})$ such that x belongs now to the cartesian product $(\mathcal{P}^* (\mathcal{F}_2))^n$. Hence we can compute $w_H(x) = card\{i \in \mathbb{N} | \ 0 \notin x_i\} = d_H(0, x)$.

The following map denoted by w_H on the cartesian product $(\mathcal{P}^* (\mathcal{F}_2))^n$:

$$w_H : (\mathcal{P}^* (\mathcal{F}_2))^n \to \mathbb{N}$$

$$a = (a_1, \dots, a_n) \mapsto card\{i \in \mathbb{N} | \ 0 \notin a_i\}.$$

is the Hamming weight on $\mathcal{F}_2^n$. So for all $x, y \in \mathcal{F}_2^n$, we have $\lceil_{\mathcal{H}}(x, y) = w_H(x \oplus y')$.

If $\mathcal{C}$ is a linear code over $\mathcal{F}_2$, the integer number $\lceil = \min\{w_H(x) | x \in \mathcal{C}\}$ is called the minimal distance of the code $\mathcal{C}$.

To characterized a linear code of length n over $\mathcal{F}_2$ as a subhypervector space of $\mathcal{F}_2^n$, it is sufficient to have a basis of that linear code. This basis can often be represented by a $k \times n$-matrix over $\mathcal{F}_2$ (where k is the dimension of the code).

We denoted by $\mathcal{M}(\mathcal{F}_2)$ be the set of all matrices over $\mathcal{F}_2$.

Definition 4.21. Let $\mathcal{C}$ be a linear code over $\mathcal{F}_2$. We called a generator matrix of $\mathcal{C}$ any matrix from $\mathcal{M}(\mathcal{F}_2)$ where the rows form a basis of the code $\mathcal{C}$.

Proposition 4.22. *Let* $\mathcal{B}_{\rfloor} \in \mathcal{M}_{k \times n}(\mathcal{F}_2)$ *be a generating matrix of the linear code* $\mathcal{C}$ *over* $\mathcal{F}_2$, *then* $\mathcal{C} = \{c \in a \cdot \mathcal{B}_{\rfloor} | \ a \in \mathcal{F}_2^k\}$.

Proof. Let $\mathcal{C}$ be a $[n,k]$-linear code over $\mathcal{F}_2$ and $\mathcal{B}_\lrcorner$ a generating matrix of $\mathcal{C}$. Then the rows of $\mathcal{B}_\lrcorner \in \mathcal{M}_{k\times n}(\mathcal{F}_2)$ form a basis of $\mathcal{C}$. So $\mathcal{C}$ consists of all linear combinations of the rows of $\mathcal{B}_\lrcorner$, therefore $\mathcal{C} = \{c \in a \cdot \mathcal{B}_\lrcorner \mid a \in \mathcal{F}_2^k\}$. □.

It is know that the dual code $\mathcal{C}^\perp$ of the linear code $\mathcal{C}$ over $\mathcal{F}_2$ is also linear, so $\mathcal{C}^\perp$ has a generating matrix called a parity check matrix.

Here and until the end of this paper, we will denoted by $\mathcal{B}_\lrcorner$ the generating matrix and by $\mathcal{H}_\lrcorner$ the parity check matrix of the linear code $\mathcal{C}$ over $\mathcal{F}_2$.

Example 4.23. *Let* $\mathcal{B}_\lrcorner = \begin{pmatrix} 1 & 0 & 1 \\ 0 & 1 & 1 \end{pmatrix}$ *be a generating matrix of the linear code C from Example 4.18. Then the parity check matrix of C is* $\mathcal{H}_\lrcorner = (1\ 1\ 1)$.

Theorem 4.24. *Let C be a linear code of length n* $(n \geq 2)$ *and dimension k over* $\mathcal{F}_2$. *Then* $\mathcal{H}_\lrcorner \in \mathcal{M}_{(n-k)\times n}(\mathcal{F}_2)$ *and* $0 \in \mathcal{B}_\lrcorner \cdot \mathcal{H}_\lrcorner^t$. *(It should be noted that* $\mathcal{H}_\lrcorner^t$ *means the transpose of* $\mathcal{H}_\lrcorner$*).*

Proof. Let the generating matrix and the parity check matrix be denoted respectively by $\mathcal{B}_\lrcorner = \begin{pmatrix} g_1 \\ \vdots \\ g_k \end{pmatrix}$ and $\mathcal{H}_\lrcorner = \begin{pmatrix} h_1 \\ \vdots \\ h_{n-k} \end{pmatrix}$, where $g_i \in \mathcal{F}_2^n$ and $h_j \in \mathcal{F}_2^n$ (for $i = 1 \cdots k$ and $j = 1 \cdots n-k$).

Then, $\mathcal{B}_\lrcorner \cdot \mathcal{H}_\lrcorner^t = \begin{pmatrix} g_1 \cdot h_1^t & g_1 \cdot h_2^t & \cdots & g_1 \cdot h_{n-k}^t \\ g_2 \cdot h_1^t & g_2 \cdot h_2^t & \cdots & g_2 \cdot h_{n-k}^t \\ \vdots & \vdots & \vdots & \vdots \\ g_k \cdot h_1^t & g_k \cdot h_2^t & \cdots & g_k \cdot h_{n-k}^t \end{pmatrix}$. Thus, by the definition of $\mathcal{C}^\perp$, $0 \in \mathcal{B}_\lrcorner \cdot \mathcal{H}_\lrcorner^t$. □

We now give some examples of linear codes over $\mathcal{F}_2$ and we make some comparison between the linear codes over the finite field with two elements $\mathbb{F}_2$ and the linear code over the Krasner hyperfield $\mathcal{F}_2$.

Example 4.25. *Let* $\mathcal{F}_2^3$ *be a hypervector space over* $\mathcal{F}_2$ *and C be a subhypervector space of* $\mathcal{F}_2^3$, *with dimensional* $k = 2$. *Then C is a linear code of length* $n = 3$ *and dimension* $k = 2$ *over* $\mathcal{F}_2$.

1. *Let* $\mathcal{B}_1 = \begin{pmatrix} 0 & 1 & 0 \\ 1 & 0 & 1 \end{pmatrix}$ *be a generating matrix of the linear code* $\mathcal{C}_1 = \{000, 010, 101, 111\}$ *over* $\mathcal{F}_2$. $\mathcal{B}_1$ *is also a generating matrix of a linear code* $\mathcal{C}_2 = \{000, 010, 101, 111\}$ *of length 3 and dimension 2 over the finite field* $\mathbb{F}_2$. *These two codes* $\mathcal{C}_1$ *and* $\mathcal{C}_2$ *have the same parameters and* $card(\mathcal{C}_1) = card(\mathcal{C}_2)$.

2. *Let* $\mathcal{B}_2 = \begin{pmatrix} 1 & 1 & 0 \\ 1 & 0 & 1 \end{pmatrix}$ *be another generating matrix of the linear code C over* $\mathcal{F}_2$. $\mathcal{B}_2$ *is also a generating matrix of a linear code* $\mathcal{C}_2'$ *of length 3 and dimension 2 over the finite field* $\mathbb{F}_2$.

Here we have that $\mathcal{C}_1 = \{000, 110, 101, 011, 111\}$, $\mathcal{C}_2' = \{000, 110, 101, 011\}$, so these two codes have the same parameters but $card(\mathcal{C}_1) > card(\mathcal{C}_2')$.

3. *Let* $\mathcal{B}_{min} = \begin{pmatrix} Id_k & Id_{n-k} \\ \cdot & 0 \end{pmatrix}$ *(where* Id_k *is the* $k \times k$*-identity matrix).*

$\mathcal{B}_{min}$ is a generating matrix of a linear code $\mathcal{C}_{min}$ of length n and dimension k over $\mathcal{F}_2$ (with $n - k \leq k$). The linear code $\mathcal{C}_{min}$ over $\mathcal{F}_2$ generated by $\mathcal{B}_{min}$ has the minimal number of code words, $card(\mathcal{C}_{min}) = 2^k$.

4. Let $\mathcal{B}_{max} = \begin{pmatrix} Id_k & 1_{n-k} \end{pmatrix}$ (where Id_k is the identity matrix and 1_{n-k} is the matrix such that every element is equal to 1).

$\mathcal{B}_{max}$ is a generating matrix of a hyperlinear code $\mathcal{C}_{max}$ of length n and dimension $k > 2$ over $\mathcal{F}_2$. The linear code $\mathcal{C}_{max}$ over $\mathcal{F}_2$ generated by $\mathcal{G}_{max}$ has the maximal number of code words, $card(\mathcal{C}_{max}) = 2^{n-k} + \sum_{i=2}^{k-1} \binom{k}{i} + k + 1$.

This remark is deduce from the previous example.

Remark 4.26. There exists a finite hyperfield such that for any other finite field of the same cardinality, the linear codes over the hyperfield are always better than the classical linear code over the finite field. (i.e., they have more code words).

In classical coding theory, one of the most important problems mentioned by MacWilliams and Sloane in their book *The Theory of Error-Correcting Codes* [27] is to find a code with a large number of words knowing the parameters (length, dimension and minimal distance). So the hyperstructure theory may help to increase the number of code words. That is the subject of the next theorem.

Theorem 4.27. *Let C be a linear code of length n and dimension k over $\mathcal{F}_2$. If M is the cardinality of C, then* $2^k \leq M \leq \begin{cases} 2^{n-k} + k + 1, & \text{if } k \leq 2; \\ 2^{n-k} + \sum\limits_{i=2}^{k-1} \binom{k}{i} + k + 1, & \text{if } k > 2. \end{cases}$

Proof. Since a generating matrix contains a basis of the linear code C as rows, it is sufficient to give a way how to construct a generator matrix for the code where the cardinality is maximal.

If $k \leq 2$, this is trivial.

If $k > 2$, then we choose a generator matrix such that:

1. in the first k columns no 1 is repeated. (this forces that every code word belongs to only one linear combination).

2. not any sum of elements in one column is equal to zero.

3. all the elements of the $n - k$ last columns are equal to 1. (because we need every combination has the maximal number of code words)

Therefore, the maximal number of code words is $2^{n-k} + \sum_{i=2}^{k-1} \binom{k}{i} + k + 1$. □

We deduce from the Theorem 4.27 what is follow, which mean that a linear code over the hyperfield F_2 satisfies the Singleton bound.

Corollary 4.28. *Let C be a linear code of length n and dimension k over $\mathcal{F}_2$, and C' be a linear code of length n and dimension k over the finite field $\mathbb{F}_2$. Then $d \leq d' \leq n - k + 1$ (where d is the minimal distance of C and d' is the minimal distance of C').*

The following next propositions give some characterization of the linear codes over $\mathcal{F}_2$ using their generating matrix and their parity check matrix.

Proposition 4.29. *Let C be a linear code of length n and dimension k over $\mathcal{F}_2$, then $c \in C$ if and only if $0 \in c \cdot \mathcal{H}_J^t$.*

Proof. $\Rightarrow$): Let $c \in \mathcal{C}$ and $\mathcal{H}_\lrcorner = \begin{pmatrix} h_1 \\ \vdots \\ h_{n-k} \end{pmatrix}$ be the parity check matrix of the code $\mathcal{C}$.
Then $c \cdot \mathcal{H}_\lrcorner{}^t = \left(c \cdot h_1^t, c \cdot h_2^t, \cdots, c \cdot h_{n-k}^t\right)$, thus by definition of $\mathcal{C}^\perp$, $0 \in c \cdot \mathcal{H}_\lrcorner{}^t$.

$\Leftarrow$) Assume that $0 \in c \cdot \mathcal{H}_\lrcorner{}^t$, then c belongs either to $\mathcal{B}_\lrcorner$, or to a linear combination of rows of $\mathcal{B}_\lrcorner$. Therefore $c \in \mathcal{C}$. □

Proposition 4.30. *Let $\mathcal{C}$ be a linear code of length n over $\mathcal{F}_2$, then the double dual of $\mathcal{C}$ is equals to $\mathcal{C}$, that is $\left(\mathcal{C}^\perp\right)^\perp = \mathcal{C}$.*

Proof. Using Proposition 4.3 in [26], $\left(\mathcal{C}^\perp\right)^\perp$ is a linear code of length n over $\mathcal{F}_2$, so it is sufficient to show that $\mathcal{C} = \left(\mathcal{C}^\perp\right)^\perp$. By definition we have $\left(\mathcal{C}^\perp\right)^\perp = \{a \in \mathcal{F}_2 |\ 0 \in y \cdot a^t;$ for all $y \in \mathcal{C}^\perp\}$, so it is straightforward that $\mathcal{C} \subseteq \left(\mathcal{C}^\perp\right)^\perp$. Now, let $a \in \left(\mathcal{C}^\perp\right)^\perp$. Let $\mathcal{H}_\lrcorner = \begin{pmatrix} h_1 \\ \vdots \\ h_{n-k} \end{pmatrix}$ be the parity check matrix of the code $\mathcal{C}$, then

$$a \cdot \mathcal{H}_\lrcorner{}^t = \left(\sum_{i=1}^n a_i \cdot h_{1,i}, \cdots, \sum_{i=1}^n a_i \cdot h_{n-k,i}\right)$$

$$= \left(\sum_{i=1}^n h_{1,i} \cdot a_i, \cdots, \sum_{i=1}^n h_{n-k,i} \cdot a_i\right) = \left(\sum_{i=1}^n h_{1,i} \cdot a^t, \cdots, \sum_{i=1}^n h_{n-k,i} \cdot a^t\right).$$

Thus $0 \in a \cdot \mathcal{H}_\lrcorner{}^t$ by definition of $\left(\mathcal{C}^\perp\right)^\perp$, therefore $a \in \mathcal{C}$. We conclude the proof by using Proposition 4.29. □

It is known from [26] that cyclic code in $\mathcal{F}_2^n$ has only one generating polynomial, so it is clear that this polynomial divides the polynomial $x^n - 1$.

Proposition 4.31. *If $g(x) = a_0 + a_1 x + \cdots + a_k x^k \in \mathcal{F}_2[x]$, is the generating polynomial for a cyclic code $\mathcal{C}$ over $\mathcal{F}_2$, then* $\mathcal{B}_\lrcorner = \begin{pmatrix} a_0 & \cdots & a_k & 0 & 0 & \cdots & 0 \\ 0 & a_0 & \cdots & a_k & 0 & \cdots & 0 \\ 0 & 0 & a_0 & & a_k & \cdots & \vdots \\ \vdots & \vdots & \ddots & \ddots & \cdots & \ddots & 0 \\ 0 & 0 & \cdots & 0 & a_0 & \cdots & a_k \end{pmatrix}$ *is the generator matrix of the cyclic code $\mathcal{C}$.*

Proof. Let $g_1 = (a_0, \ldots, a_k, 0, \ldots, 0) \in \mathcal{F}_2^n$, then $\mathcal{B}_\lrcorner$ can also be write as

$\mathcal{B}_\lrcorner = \begin{pmatrix} g_1 \\ s(g_1) = g_2 \\ s^2(g_1) = g_3 \\ \vdots \\ s^{k-1}(g_1) = g_k \end{pmatrix}$ (where s is the shift function and $s^k = s \circ s \circ \cdots \circ s$, k-successive shifts).

Since the polynomial g generates $\mathcal{C}$, we have $\mathcal{C} = \langle g(x) \rangle$. Let $c \in \mathcal{C}$, then $(c_i)_{i=1 \cdots n} = c \in g(x) \cdot p(x)$ (where $b_0 + b_1 x + \cdots + b_{n-1} x^{n-1} = p(x) \in \frac{\mathcal{F}_2[x]}{(x^n - 1)}$) implies that $c_i \in \sum_{l+j} a_l \cdot b_j$ if $i \leq k$ and $c_i = 0$ else if $(i > k)$.

Focus on $g(x)$ and $p(x)$, the element c belongs to the sum $b_0 \cdot g(x) + b_1 x \cdot g(x) + \cdots + b_{n-1} \cdot x^{n-1} \cdot g(x)$ because this sum can also be written as $e_1 \cdot g_1 + e_2 \cdot g_2 + \cdots + e_k \cdot g_k$ $(e = (e_1, \ldots, e_k) \in \mathcal{F}_2^n)$, and $\mathcal{C}$ is a cyclic code generated by $g(x)$. □

The following Proposition use same notations as in Proposition 4.31.

Proposition 4.32. *[28] Let* $h(x) \in \frac{\mathcal{F}_2[x]}{(x^n-1)}$ *be a polynomial such that* $x^n - 1 \in h(x) \cdot g(x)$, *then*

1. *The linear code* C *over* $\mathcal{F}_2$ *can be represented as* $C = \left\{p(x) \in \frac{\mathcal{F}_2[x]}{(x^n-1)} | \ 0 \in p(x) \cdot h(x)\right\}$.

2. $h(x)$ *is the generating polynomial for the linear code* $C^{\perp}$.

To illustrate what is doing for cyclic codes and polynomials, we have this example.

Example 4.33. *Let* C *be a linear code over* $\mathcal{F}_2$ *generate by the polynomial* $g(x) = 1 + x^2 \in \frac{\mathcal{F}_2[x]}{(x^3-1)}$. *Then the generator matrix of the code* C *is given by* $\mathcal{B}_{\lrcorner} = \begin{pmatrix} 1 & 0 & 1 \\ 1 & 1 & 0 \end{pmatrix}$.

Since $x^3 - 1 \in (1 + x^2)\dot{\odot}\}(1 + x + x^2)$, then the polynomial $h(x) = 1 + x + x^2$ is the parity check polynomial of the code C, and the parity check matrix is given by $\mathcal{H}_{\lrcorner} = (1 \quad 1 \quad 1)$.

Thus $C = \left\{p(x) \in \frac{F_2[x]}{(x^3-1)} | \ x^3 - 1 \in p(x)\dot{\odot}\}(1 + x + x^2)\right\} = \left\{0, 1 + x^2, 1 + x, 1 + x + x^2, x + x^2\right\}$.

5. Conclusions

This Chapter divides in three sections **Fuzzy linear codes over** $\mathbb{Z}_{p^k}$, **Fuzzy** $\mathbb{Z}_{p^k}$**-linear codes** and **Linear codes over Krasner hyperfields** just introduce some new perspectives in the field of coding theory. In the first and second section, we define and give some related properties of these on codes. We show in some example that fuzzy linear code can deal with uncertain information directly. The third section, which joint the previous sections in the sense that fuzzy fields/rings and Krasner hyperfields are non classical structures which approxim very well many real life situation, study linear codes over Krasner hyperfields as linear codes over finite fields. Many of their properties are given and the important thing that arise here is that with almost the same parameters linear codes construct on Krasner hyperfields have much code words than one construct on fields.

Author details

Surdive Atamewoue Tsafack
University of Yaounde I, Yaounde, Cameroon

*Address all correspondence to: surdive@yahoo.fr

References

[1] L.A. Zadeh, Fuzzy sets, *Information and Control* **8** 338-353 (1965).

[2] F. Marty, Sur une generalization de la notion de groupe, 8^{iem} *congres Math. Scandinaves,Stockholm*, 45-49 (1934).

[3] M. Krasner, A Class of Hyperrings and Hyperfields, *Internat. J. Math. and Math. Sci.* **6**, 307-312 (1983).

[4] C.E. Shannon, Communication in presence of noise, *IEEE*, **37**, 10-21 (1949).

[5] K. P. Shum, Chen De Gang, Some note on the theory of fuzzy code, *Electronic BUSEFAL-81, Polytech.univ-savoie, France* 132-136 (2000).

[6] L. O. Hall and G. Diall, On fuzzy code for asymmetric and unidirectional errors, *Fuzzy sets and systems* **36**, 365-373 (1990).

[7] P.A. Von Kaenel, Fuzzy codes and distance properties, *Fuzzy sets and systems* **8**, 199-204 (1982).

[8] F. Galand, Construction de codes $\mathbb{Z}_{p^k}$ -linéaires de bonne distance minimale et schémas de dissimulation fondés sur les codes de recouvrement, Ph.D Thesis, Université de Caen, (2004).

[9] M. Maschinchi and M.M. Zahedi, On L-fuzzy primary submodules, *Fuzzy Sets and Systems* **49**, 231-236 (1992).

[10] C.V. Negoita and D.A. Ralescu, Applications of Fuzzy Sets and System Analysis, (*Birkhous, Basel*) (1975).

[11] R. Biswas, Fuzzy fields and fuzzy linear space redefined, *Fuzzy Sets and Systems* **33**, 257-259 (1989).

[12] *S. Nanda*, Fuzzy fields and fuzzy linear space, *Fuzzy Sets and Systems* **19**, 89-94 (1986).

[13] M. Kondo, Wieslaw A. Dubek, On transfer principle in fuzzy Theory, *Mathware and soft computing* **12**, 41-55 (2005).

[14] C. Carlet, $\mathbb{Z}_{2^k}$-linear codes, *IEEE Transactions on Informations Theory* **44**, 1543-1547 (1998).

[15] S. Atamewoue Tsafack , S. Ndjeya , L. Strüngmann and C. Lele, Fuzzy Linear Codes, *Fuzzy Information and Engineering*, https://doi.org/10.1080/16168658.2019.1706959 (2020).

[16] L.A. Zadeh, The concept of a linguistic variable and its application to approximate reasoning I, II, III, *Information Sciences* **8-9**, 199-257, 301-357, 43-80 (1975).

[17] A.M. Kerdock, A class of low-rate nonlinear codes, *Information and Control* **20** (1972).

[18] R. Hammons, P.V. Kumar, A.R. Calderbank, N.J.A. Sloane et P. Solé, Kerdock, Preparata, Goethals and other codes are linear over $\mathbb{Z}_4$, *IEEE Transactions on Information Theory* **40**, 301-319 (1994).

[19] I. Perfilieva, Fuzzy function: Theoretical and practical point of view. *Atlantis Press* 480-486 (2011).

[20] R. Ameri and O.R. Dehghan, On Dimension of Hypervector Spaces, *European Journal of Pure and Applied Mathematics* **1**, 32-50 (2008).

[21] P. Corsini and V. Leoreanu, Applications of Hyperstructure Theory, *Kluwer Academical Publications, Dordrecht*, (2003).

[22] B. Davvaz and V. Leoreanu-Fotea, Hyperring Theory and applications, International Academic Press, USA, (2007).

[23] Sanjay Roy and T.K. Samanta, A Note on Hypervector Spaces, *Discussiones MathematicaeGeneral Algebra and Applications* **31**, 75-99 (2011).

[24] S. Jančic-Rašović, About the hyperring of polynomial, *Ital. J. Pure Appl. Math.* **21**, 223-234 (2007).

[25] F. Galand, Construction de codes $\mathbb{Z}_{p^k}$-linéaires de bonne distance minimale et schémas de dissimulation fondés sur les codes de recouvrement, Ph.D Thesis, Université de Caen, (2004).

[26] B. Davvaz and T. Musavi, Codes Over Hyperrings, *Matematički Vesnik* **68**, 26-38 (2016).

[27] F.J. MacWilliams and N.J.A. Sloane, The Theory of Error-Correcting Codes, *North-Holland, Amsterdam*, (1977).

[28] S. Atamewoue Tsafack, S. Ndjeya, L. Strüngmann and C. Lele, Codes over Hyperfields, Discussioness Mathematicae Genenal Algerbra and Applcations **37**, 147-160 (2017).